MANAGING TOURIST HEALTH AND SAFETY IN THE NEW MILLENNIUM

ADVANCES IN TOURISM RESEARCH

Series Editor: Professor Stephen J. Page, Scottish Enterprise Forth Valley Chair in Tourism,
Department of Marketing, University of Stirling, Scotland, U.K.
s.j.page@stir.ac.uk

Advances in Tourism Research series publishes monographs and edited volumes that comprise state-of-the-art research findings, written and edited by leading researchers working in the wider field of tourism studies. The series has been designed to provide a cutting edge focus for researchers interested in tourism, particularly the management issues now facing decision-makers, policy analysts and the public sector. The audience is much wider than just academics and each book seeks to make a significant contribution to the literature in the field of study by not only reviewing the state of knowledge relating to each topic but also questioning some of the prevailing assumptions and research paradigms which currently exist in tourism research. The series also aims to provide a platform for further studies in each area by highlighting key research agendas which will stimulate further debate and interest in the expanding area of tourism research. The series is always willing to consider new ideas for innovative and scholarly books, inquiries should be made directly to the Series Editor.

Published:

BAUM & LUNDTORP
Seasonality in Tourism

ASHWORTH & TUNBRIDGE
The Tourist-Historic City: Retrospect and Prospect of Managing the Heritage City

RYAN & PAGE
Tourism Management: Towards the New Millennium

SONG & WITT
Tourism Demand Modelling and Forecasting: Modern Econometric Approaches

TEO, CHANG & HO
Interconnected Worlds: Tourism in Southeast Asia

Forthcoming titles include:

KERR
Tourism Public Policy and the Strategic Management of Failure

THOMAS
Small Firms in Tourism: International Perspectives

PAGE & LUMSDON
Progress in Tourism and Transport Research

Related Elsevier Journals — sample copies available on request

Annals of Tourism Research
Cornell Hotel and Restaurant Administration Quarterly
International Journal of Hospitality Management
International Journal of Intercultural Relations
Tourism Management
World Development

MANAGING TOURIST HEALTH AND SAFETY IN THE NEW MILLENNIUM

EDITED BY

JEFF WILKS

The University of Queensland, Australia

STEPHEN J. PAGE

University of Stirling, Scotland, U.K.

2003

Pergamon
An Imprint of Elsevier Science

Amsterdam – Boston – London – New York – Oxford – Paris
San Diego – San Francisco – Singapore – Sydney – Tokyo

ELSEVIER SCIENCE Ltd
The Boulevard, Langford Lane
Kidlington, Oxford OX5 1GB, UK

First edition 2003

Library of Congress Cataloging in Publication Data
A catalog record from the Library of Congress has been applied for.

British Library Cataloguing in Publication Data
A catalogue record from the British Library has been applied for.

ISBN: 0-08-044000-2

All chapters in this text were independently peer reviewed prior to acceptance.

Cover according to a design by Ian Hutson and Pam Koger, Brisbane, Australia.
Cover photographs courtesy of Tourism Queensland.

♾ The paper used in this publication meets the requirements of ANSI/NISO Z39.48-1992 (Permanence of Paper).
Printed in The Netherlands.

Contents

Section 3: Advice and Best Practice

Section 4: Selected Issues

Conclusions

List of Figures

List of Tables

List of Plates

Contributors

Trevor Atherton
Trevor Atherton is an international consultant on tourism planning, development, policy and law. He is currently Senior Advisor and Program Manager for the Supreme Commission for Tourism in Saudi Arabia.

Trudie Atherton
Trudie Atherton is a Lecturer at the University of Technology, Sydney and visiting Associate Professor at Bond University, Queensland. She is a practising lawyer with a particular interest in mediation, training and tourism awareness.

Michael Barker
Michael Barker was a Research Fellow in Tourism at Massey University, Auckland from 1999–2000.

Tim Bentley
Tim Bentley was a Research Fellow in Tourism at Massey University during 1999 and then worked at Forest Research in Rotorua. He now lectures at Massey University, Auckland.

Alison Blackwell
Alison Blackwell is Senior Research Fellow at the University of Edinburgh, Scotland. She is an applied entomologist with an interest in biting midges, their biology and intervention strategies for their control.

Johnathan Cossar
Johnathan Cossar is a general medical practitioner in Glasgow and Research Associate at the Scottish Centre for Infection and Environmental Health.

Alan Ewert
Alan Ewert is a Professor and the Patricia and Joel Meier Endowed Chair of Outdoor Leadership at Indiana University, Bloomington, United States of America.

Leisa Holzheimer
Leisa Holzheimer is Research Officer with the Centre for Tourism and Risk Management at The University of Queensland, Australia.

Lynn Jamieson
Lynn Jamieson is Associate Professor and Chair in the Department of Recreation and Park Administration at Indiana University, Bloomington, United States of America.

Ian Laird
Ian Laird is Senior Lecturer in Safety Management with the Department of Human Resource Management at Massey University, Palmerston North, New Zealand.

Denny Meyer
Denny Meyer is Associate Professor in Statistics at Massey University, Auckland. She is a specialist in multivariate techniques and their application to tourism modelling and forecasting.

Stephen J. Page
Stephen Page is Scottish Enterprise Forth Valley Chair in Tourism and Professor of Tourism Management at the University of Stirling. Prior to this appointment, he was Professor of Tourism Management at Massey University, New Zealand.

Donna Pendergast
Donna Pendergast is Principal Research Fellow with the Centre for Tourism and Risk Management and lectures in education at The University of Queensland, Australia.

Bruce Prideaux
Bruce Prideaux is Senior Lecturer in the School of Tourism and Leisure Management at The University of Queensland. He has an extensive background in the tourism and transport industries.

Chris Ryan
Chris Ryan is Professor of Tourism at the University of Waikato, New Zealand. He is editor of the journal 'Tourism Management' and a member of the International Academy for the Study of Tourism.

Linda Walker
Linda Walker is a doctoral candidate examining the management of tourist health and safety in the department of Marketing, University of Stirling. Her research was conducted in conjunction with Central Scotland Police.

Jeff Wilks
Jeff Wilks is Professor of Tourism and Director of the Centre for Tourism and Risk Management at The University of Queensland, Australia. A qualified psychologist and lawyer, he has a special interest in the health and safety of tourists.

Dedication

This book is dedicated to the memory of William (Bill) Faulkner, Professor of Tourism Management at Griffith University, Queensland, Australia (1993–2002).

Bill Faulkner was an exceptional scholar with a particular interest in tourist safety. One of his papers, *Towards a Framework for Tourism Disaster Management* is reproduced in Chapter 11 from the original that appeared in the journal *Tourism Management*.

Until just prior to his death in January 2002, Professor Bill Faulkner was the Deputy CEO and Director of Research in the Cooperative Research Centre for Sustainable Tourism, and Professor of Tourism Management at Griffith University, Australia.

He was also a Fellow and Member of the Board of Directors of the Australian Tourism Research Institute (ATRI) and served as a member of the Editorial Boards of several international journals (*Tourism Management, Tourism Analysis, Journal of Tourism Studies, Pacific Tourism Review, Journal of Quality Assurance in Tourism and Hospitality, Event Management, Turizam* and *Anatolia*).

Prior to taking up these academic positions, Bill was the founding Director of Australia's Bureau of Tourism Research and was instrumental in setting up the Australian Federal Government's Tourism Forecasting Council.

Foreword

Overseas travel has always held an element of risk. For the early missionaries and explorers to developing countries that risk included exotic diseases, unpredictable weather, attacks by wild animals, and in some cases, conflict with local residents whose background and customs were totally different to those of the visitors.

Early travellers managed their health and safety risks based on available knowledge and technology. That was part of the adventure of travel. Today, travel still contains elements of adventure and discovery, though some of the risks to be managed are quite different. For example, the terrorist attacks of 11 September 2001 in the United States, and the more recent bombing of tourist facilities in Bali, have shown that personal safety and security in public places is a key issue for governments, the tourism industry and individual travellers as we move into the new millennium.

This book is a valuable and timely contribution to our understanding of current issues in the broad area of tourist health and safety. Like the early explorers, we rely on available knowledge and technology in order to effectively manage risk. In addition to medical conditions and injuries that affect travellers, the book explores emerging issues such as the adventure tourism market, travel and tourism law, the role of travel agents in providing health and safety advice, and best practice requirements at both a business level and for destinations as a whole.

This book also shows how applied research can be used to identify and respond to risk — providing current and accurate data to guide management decisions. The overall theme of tourist well-being is very appropriate, since we have now entered an era where protecting the health of tourists, and ensuring their general safety in a range of leisure activities, is very much part of delivering a quality service at all tourist destinations.

I congratulate the editors and contributors on a book that focuses our attention on tourist health and safety as a key management issue for the future. My friend and colleague Bill Faulkner was a pioneer in researching destination management issues. This text is a fitting tribute to his memory.

Sir Frank Moore
Chair, Australian Tourism Forecasting Council
Chair, Cooperative Research Centre for Sustainable Tourism

Acknowledgments

The editors would like to thank Ms Leisa Holzheimer for managing this project. Special thanks also to Dr Donna Pendergast (The University of Queensland, Australia) for reviewing the submitted chapters.

The following provided copyright permission to reproduce material: Mr G. Hendry for Figure 12.1, redrawn from his original of 'The Distribution of Midges in Scotland'; Mr John Brown, Scottish Executive, for permission to reproduce Figure 12.4, The Tourism Framework for Action 2002–2005; Mr Paul Broughton for permission to reproduce Figure 14.4, A Risk Assessment Program. Tom Clark and Neil Boon at Elsevier Science have been very helpful in steering the book through the various stages through to completion.

Introduction

Chapter 1

Current Status of Tourist Health and Safety

Jeff Wilks and Stephen J. Page

The terrorist attacks in the United States of America on 11 September 2001 changed forever our views of traveller safety and security. The impact of these attacks has been far-reaching and has seriously affected the way the tourism industry operates. One month after the event, the International Labor Organization (ILO) estimated that nine million jobs would be lost in the tourism industry as a direct result of these attacks and predicted that it will take years for employment to return to pre-11 September levels (www.ilo.org/public/english/bureau/inf/pr/2001/36.htm).

While 11 September 2001 focused world attention on terrorism, it is important to appreciate that tourist health, safety and security issues have been quietly gaining recognition in the background during the past decade. 'In the background' is the qualifier, as tourism is about selling dreams. Irrespective of how important health and safety issues become, some tourism-marketing officials will always play them down since 'bad news' in any form is not conducive to selling holidays (Wilks & Oldenburg 1995).

This attitude toward tourist health and safety is slowly changing as public liability insurance becomes more difficult to obtain and insurers are requiring operators to minimise their exposure to claims by adopting 'best practice' risk management strategies (Liability Insurance Taskforce 2002). Best practice strategies call for an acceptance that things may go wrong; that customers are sometimes injured; that in contract law promises about a holiday must be kept; and that regular staff training and operations monitoring are a necessary component of modern business. In the United Kingdom, recent research by the Association of British Travel Agents (ABTA) estimated that around 5% of the 20 million annual package holidays taken overseas had customers who were fairly or very dissatisfied. Health, safety and security issues certainly feature as one element in the tourist dissatisfaction equation (Anonymous 2001). However, tourist health and safety is no longer just a burden for the industry. In many cases good health and safety practices generate income (Caribbean HEAT www.carec.org/projects/hotels/carib_heat.htm; Hospedales 1997) and at the very least, good practices will prevent lawsuits, disappointment, financial cost and lost customers.

Managing Tourist Health and Safety in the New Millennium
Copyright © 2003 by Elsevier Science Ltd.
All rights of reproduction in any form reserved.
ISBN: 0-08-044000-2

This book explores the developing topic of tourist health and safety at the beginning of the new millennium. Throughout the book the general term 'health, safety and security in tourism' refers to the protection of life, health and the physical, psychological and economic integrity of travellers, tourism employees and host communities. It also implies the safeguarding of the security interests of tourism entrepreneurs and operators and the countries sending and receiving visitors. This is the risk identification approach adopted by the World Tourism Organization (2003).

Sources of Health, Safety and Security Risk in Tourism

According to the World Tourism Organization (2003), risks to the safety and security of tourists, host communities and tourism employees originate from the following four sources:

- The human and institutional environment outside the tourism sector;
- The tourism sector and related commercial sectors;
- Individual travellers (personal risks); and
- Physical or environmental risks (natural, climatic, and epidemic).

It is worth exploring these to illustrate the full extent of potential risks in each sector.

The Human and Institutional Environment

The risks from the human and institutional environment exist when visitors fall victim to:

- Common delinquency (e.g. theft, pick-pocketing, assault, burglary, fraud, deception);
- Indiscriminate and targeted violence (e.g. rape) and harassment;
- Organized crime (e.g. extortion, slave trade, coercion);
- Terrorism and unlawful interference (e.g. attacks against state institutions and the vital interests of the state), hijacking and hostage taking;
- Wars, social conflicts and political and religious unrest; and
- A lack of public and institutional protection services.

Risks that occur in the broader community impact similarly on tourists and residents. Tourists are not always targeted, but are often caught up in events by being in the wrong place at the wrong time. Protection of tourists at this level is the responsibility of national governments and contributes to whether a destination is perceived to be safe.

Tourism and Related Sectors

Through defective operation, tourism and sectors related to tourism such as transport, sports, and retail trade can endanger visitors' personal security, physical integrity and economic interests through:

- Poor safety standards in tourism establishments (e.g. fire, construction errors, lack of anti-seismic protection);
- Poor sanitation and disrespect for the environment's sustainability;
- The absence of protection against unlawful interference, crime and delinquency at tourism facilities;
- Fraud in commercial transactions;
- Non-compliance with contracts; and
- Strikes by staff.

The protection of tourists from problems occurring in areas directly related to tourism is the joint responsibility of tourism authorities at each destination, tourism industry associations and relevant sectors of local government. Problems in this area are not necessarily the 'fault' of the tourism industry, but can have a dramatic and negative effect on a destination's image. For example, the Childers backpacker fire in Queensland during 2000 resulted in the deaths of 12 overseas visitors (Queensland Fire and Rescue Authority 2000). A jury found Robert Paul Long guilty on 18 March 2002 of two counts of murder and one count of arson. Even though Long was not in any way associated with the tourism industry, the fact that tourists were killed in the fire focused international media attention on Queensland as a travel destination. The fire also drew attention to widespread problems of fire safety within Queensland budget accommodation, resulting in a massive government audit of all facilities throughout the State.

Individual Travellers

Travellers or visitors can endanger their own safety and security, and those of their hosts by:

- Practicing dangerous sports and leisure activities, dangerous driving, and consuming unsafe food and drink;
- Travelling when in poor health, which may deteriorate during the trip;
- Causing conflict and friction with local residents, through inappropriate behaviour toward local communities or by breaking local laws;
- Carrying out illicit or criminal activities (e.g. trafficking in illicit drugs);
- Visiting dangerous areas; and
- Losing personal effects, documents and money through carelessness.

Most travel health and safety problems occur at the level of individual travellers. Studies of insurance claims, for example, show that minor illness, lost luggage and theft predominate (Ryan 1996; Leggat & Leggat 2002). Motor vehicle crashes remain the leading cause of injury-related death for tourists' worldwide (Wilks 1999; Page *et al.* 2001), followed by drowning (Wilks 2003). Unfamiliar adventure activities, such as scuba diving, account for a significant number of tourist hospital admissions each year (Wilks & Coory 2000; 2002), while pre-existing illness continues to be the main cause of fatalities (Wilks, Pendergast & Wood 2002).

For many years there was an expectation, probably stemming from the tourist operator's legal 'duty of care' responsibilities (see Atherton & Atherton 1998) that travellers should be protected from the consequences of their own actions. This seems to be changing. While operators still have a duty to warn customers of things that are unfamiliar and perhaps unexpected, the onus of taking responsibility for yourself and your actions seems to have, at least to some extent, moved back to the individual traveller. Travellers now have access to a vast knowledge base about travel issues and therefore they have fewer excuses for being uninformed or unprepared for travel. At the same time, people are looking for vacation experiences that are very different to their day-to-day routines. Consequently, there is currently an enormous growth in the area of adventure travel. Later chapters specifically examine the health and safety issues that arise from adventure travel, and the role and responsibilities of both operators and participants.

Physical and Environmental Risks

Finally, physical and environmental damage can occur if travellers:

- Are unaware of the natural characteristics of the destination, in particular its flora and fauna;
- Are not medically prepared for the trip (vaccinations, prophylaxis);
- Do not take the necessary precautions when consuming food or drink or for their hygiene; and
- Are exposed to dangerous situations arising from the physical environment (e.g. natural disasters and epidemics).

Physical and environmental risks are also largely personal risks, but are not caused deliberately. Rather, these result from the traveller's ignorance or his/her disregard for potential risks. While physical and environmental risks do not feature prominently in the tourist health and safety literature, a single environmental disaster has the potential to claim a large number of lives (World Tourism Organization and World Meteorological Organization 1998).

Since the terrorist attacks of 11 September 2001 in the United States, the term 'risk management' has become strongly associated with the physical safety and security of travellers. Surveys show that consumers list safety and security as a very high priority (Taylor 2001) and businesses have needed to respond accordingly. While there are many types of risk to which the tourism industry must respond across the four broad areas identified earlier, the focus of the new millennium centres on the physical safety of customers and staff.

The chapters in this book reflect the current interest in physical safety, especially prevention of injuries in adventure tourism activities. As travellers worldwide become more interested and involved in ecotourism, personal health promotion, outdoor activities and travel to remote destinations, the growth of adventure tourism will continue. Given that the research shows international tourists are most likely to be injured or killed while in unfamiliar environments and while participating in unfamiliar activities (Page & Meyer 1996), the focus on adventure tourism is timely and appropriate.

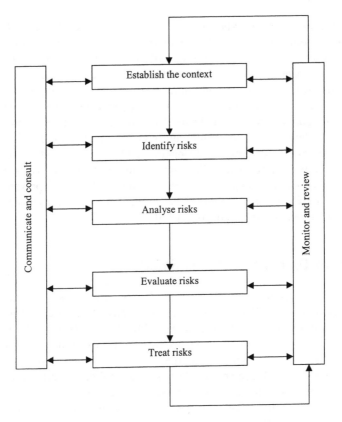

Figure 1.1: Risk management overview.

Risk Management

Any current discussion about the health and safety of tourists must include an appreciation of risk management as a concept, and the steps required to systematically address potential problem areas confronting most modern businesses. A national standard for risk management has been available in Australia and New Zealand since 1995 (Standards Australia and Standards New Zealand 1999). The standard provides a step-by-step framework for taking control of risks and their impacts. The basic framework is presented in Figure 1.1 and is well suited to managing the risks associated with tourist health, safety and security.

Identifying Risks

Most of the available literature on tourist health and safety is at the level of identifying risk, and to a lesser extent, understanding its causes and consequences. However, there are still significant gaps in our knowledge, especially across destinations. One approach to identifying health and safety risks for tourists is to consider 'snapshots' of visitor accidents at various levels. Wilks (2003) describes four levels: deaths; serious injury and illness requiring hospitalisation; injuries and illness treated by emergency departments and general medical practitioners; and other health areas, including treatment, support and advice provided for tourists by paramedic and emergency services, the Coast Guard, pharmacists, lifeguards and local citizens. In Australia, Surf Life Saving Queensland provides an excellent example of a community-based organization playing a major role in tourism and injury prevention. During the 2000–2001 season, Surf Life Saving Queensland members performed 3,370 rescues and 42 resuscitations, provided 14,964 first aid and 9,176 marine stinger treatments, and initiated 152,578 preventative actions (Surf Life Saving Queensland 2001). These services were acknowledged during 2001 with state and national tourism industry awards.

Snapshots of Tourist Health and Safety

A review of the literature reveals that only a handful of key studies have been undertaken on tourist fatalities. These include deaths of American (Hargarten *et al.* 1991), Scottish (Paixao *et al.* 1991), Canadian (MacPherson *et al.* 2000) and Australian (Prociv 1995) citizens travelling overseas. More recently, Wilks and his colleagues (Wilks *et al.* 2002) have reported on international visitors who died while in Australia over a four-year period. These studies show that natural causes are the leading reason for fatalities, particularly pre-existing illnesses such as cardiovascular disease. Motor vehicle crashes and drowning are the leading causes of injury-related deaths. Unfortunately, as the collection of information on travellers' deaths relies heavily on local police and coroners' reports, and cooperation by consular authorities, fatality studies are expensive, time consuming and difficult to conduct. However, these studies provide a very good measure of the overall safety of a destination, especially the extent that homicide, suicide, infectious disease and other preventable causes of death involve visitors.

At the level of serious illness and injury requiring admission to hospital, more information is becoming available on tourists. Again, pre-existing illness and medical conditions such as diarrhoea are the main reasons for hospital admission (Nicol *et al.* 1996). Motor vehicle crashes and drowning are the leading causes of injury-related admissions, though studies in coastal and marine-based destinations show a substantial number of tourist injuries from scuba diving, snorkelling and other water-related activities (Nicol *et al.* 1996; Wilks & Coory 2002). Because most destinations around the world have hospitals that record patient information in a standard format known as The International Classification of Diseases (World Health Organization 1978) hospital records are one of the most promising, yet underdeveloped, sources of knowledge on tourist health and safety.

The other two levels or 'snapshots' of tourist health and safety treatment for illness and injury at health clinics, and by general medical practitioners, lifeguards and other health and emergency service providers are still very underdeveloped. Reports from clinics on tropical islands (e.g. Wilks *et al.* 1995) show that tourists present with a range of pre-existing medical conditions (e.g. heart disease, respiratory problems), as well as many injuries that result from being in an unfamiliar environment (e.g. sunburn, bites and stings) and participating in unfamiliar activities (e.g. coral cuts, decompression illness from scuba diving).

The risk management approach to tourist health and safety highlights the importance of identifying sources of risk as a first step, and then using this information to understand the causes and consequences of the risk. The contribution of scholarly research is very important to this process, and it is only in fairly recent times that academics have made a substantial contribution to what is still a very new field of study. In order to fully address risk issues it is necessary to form active partnerships involving government agencies, industry groups, academics and key community stakeholders at each travel destination. Again, this is the approach adopted by the World Tourism Organization (1997; 2003).

Partnerships for Tourist Health and Safety

The World Tourism Organization is the leading international organization in the field of travel and tourism and is the United Nation's tourism agency, entrusted with the promotion and development of tourism. The World Tourism Organization serves as a global forum for tourism policy issues and is a practical source of tourism know-how. As noted earlier, tourist health and safety issues have been quietly developing over a number of years. The World Tourism Organization has been very active in this development, with the following among its key initiatives:

1989   Decision of the Executive Council to increase activities in safety and security
1991   *Recommended Measures for Tourism Safety* (booklet)
1991   *Travellers' Health Abroad* (booklet)
1991   *Creating Tourism Opportunities for Handicapped People in the Nineties* (book)
1993   *Health Information and Formalities in International Travel* (book)
1993   *Sustainable Tourism Development: A Guide For Local Planners* (book)
1994   Experts Meeting on Tourist Safety and Security, Madrid
1995   *Handbook on Natural Disaster Reduction in Tourist Areas* (book)
1995   *Indicators for Tourism and the Environment* (book)
1996   *Tourist Safety and Security: Practical Measures for Destinations* (book, first edition)
1997   Africa conference
1997   Warsaw seminar
1997   *Tourist Safety and Security: Practical Measures for Destinations* (book, second edition)
1998   ESCAP seminar
1998   Russia Far East seminar
1998   Central American seminar

1999   ECOWAS seminar
1999   Middle East seminar
2001   September 11th Crisis Committee Meeting at World Travel Market in London
2002   September 11th Recovery Committee Meeting at ITB in Berlin

The most recent contribution by the World Tourism Organization (2003) is a new manual *Safety and Security in Tourism: Partnerships and Practical Guidelines for Destinations*. The manual presents strategies for forming partnerships across government agencies and tourism industry groups to address the full range of tourist health and safety issues. It also describes successful partnerships using international case studies. The predominant message of the manual is that protecting and promoting the health and safety of tourists is an essential element of delivering quality service, and a part of the total tourist experience that destinations cannot afford to ignore.

In terms of the Risk Management model (Figure 1.1), the World Tourism Organization has been very active in all levels of analysis and evaluation of risk, as well as providing advice to its Member States on strategic management responses, such as working with the media in times of crisis. Since 1989 the World Tourism Organization has contributed substantially to this field through monitoring and review of new material and approaches to risk management.

Other Initiatives in Tourist Health and Safety

In recent years, a number of tourism industry organisations have focused their attention on tourist health and safety issues, especially the responsibilities of their members. Examples include an excellent handbook on health and safety produced by the Federation of Tour Operators (1999) in the United Kingdom, which is complemented by industry training and education seminars offered throughout Europe; major international meetings on health, safety and security issues by the International Hotel and Restaurant Association (1998; 2000); and ongoing safety and security development for the aviation industry by the International Air Transport Association (IATA).

There has also been a large increase in academic literature, with edited tourism books by Clift & Page (1996), Clift & Grabowski (1997), and Pizam & Mansfeld (1996). Separate but complementary literature has developed in travel medicine, with detailed textbooks by DuPont & Steffen (1997) and Keystone *et al.* (2003). While travel medicine has traditionally been most interested in infectious diseases, an increasing number of articles on tourist injuries and general travel safety topics are starting to appear in the *Journal of Travel Medicine*.

While the above initiatives are very positive, it is important to keep in mind that tourism is the world's largest industry (World Tourism Organization 2002) and that health and safety issues are only on the industry's agenda because of the current economic and business environment. The terrorist attacks of 11 September 2001 focused customer attention on personal safety and the tourism industry has had to respond by way of reassurance. At the same time, legal challenges and particularly the global insurance crisis have forced tourist operators to critically assess their risk environment.

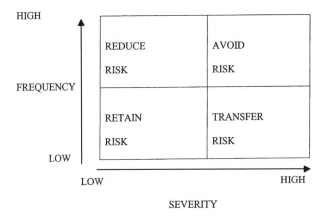

Figure 1.2: The risk evaluation matrix.

Insurance Issues

For tourism operators, strategies for managing risk fall into one of four broad categories, depending on the likely frequency of risks occurring and their severity of impact. Generally, the options include avoiding, transferring, mitigating or accepting risk (see Figure 1.2; Wilks & Davis 2000).

Where the severity or consequences of risk are high (e.g. a customer being injured or killed; loss of property in a storm or cyclone; the potential for a major lawsuit for negligence) then tourist operators traditionally relied on insurance to transfer their risk to a third party. With the current crisis in the global insurance industry (see Trowbridge Consulting 2002) many operators are unable to obtain insurance cover, or alternatively are paying large premiums for their cover.

One of the main factors in the escalating costs is personal injury and the large compensations injured parties are receiving through the court system (Liability Insurance Taskforce 2002). In order to address this issue operators need to shift their attention to reducing risk through 'best practice' initiatives such as having written policies and procedures, staff training, signage, visitor and customer briefings, and monitoring of industry standards (Department of Industry, Tourism & Resources 2002).

Without effective risk management practices, tourist operators will continue to struggle with health and safety issues, both financially and legally. At the same time, customers now have increased expectations that their health and safety will be protected. Destinations that can establish and maintain a reputation for safety have a very competitive advantage as we move into the new millennium. Conversely, those travel destinations that are perceived as having health and safety problems will lose customers (see Wanderlust Magazine 2001).

The chapters in this book also raise the spectre of new research agendas and conceptualisations of tourist health and safety that may be broadly labelled 'tourist well-being.' This theme runs throughout the book, and for this reason, it is pertinent to consider its significance before turning to the individual chapters of the book.

Tourist Well-Being: A Critical Concept in Managing Tourist Health and Safety

Ensuring that tourist well-being is safeguarded in destinations where tourist activities involve a high degree of risk, requires a management process where the excitement and challenge posed by risk behaviours are balanced with appropriate safety measures and organisational systems. It is inappropriate in an era of growing litigation to place the emphasis on the consumer, since they may not be adept at assessing if appropriate safety systems are in place for activities involving risk. One of the problems of developing a wider conceptualisation of tourist safety to incorporate notions of well-being means moving towards a more holistic assessment where a new research paradigm based on well-being is necessary. This places the tourist as the focal point of the research and requires a wide range of multidisciplinary skills to assess the wider context of safety issues and how these impact on tourists' well-being.

The concept of well-being has been recognised and used within human geography for over two decades (see Smith's 1977 seminal study in the area). Smith's (1997) study highlights the usefulness of a human welfare geography perspective in understanding human well-being and the factors which impact upon it in different contexts (e.g. the leisure environment). A well-being approach requires the contribution of different disciplines to be synthesised to contribute to analyses about specific tourist outcomes in time and space. In a tourism context, this perspective can be applied to understand how a customer's well-being is conditioned and affected by what they do, where they stay and the risks to which they are exposed. In other words, one needs to understand the interaction of the tourist and tourism industry in a particular locale and how safety issues impact upon their well-being. For example, a tourism perspective will assist in understanding the context of the tourist visit, motivation for the trip and their preferred activity patterns.

In geographically informed research (Hall & Page 2002) one can understand why tourists do what they do and where (e.g. adventure tourism) and thereby the impacts they generate, particularly the incidence of adverse experiences. Research on outdoor recreation can help one to understand the factors which affect the decision-making of tourists to choose activities in the natural environment that involve physical risk such as adventure tourism. In health-related research that is informed by the discipline of psychology, it is possible to uncover the tourists' cognition of warnings and risk associated with adventure. Safety management research can also assist in assessing risk and in the investigation of accident causation to assess what chain of events contributed to specific tourist accidents. In addition, the contribution of the statisticians' skills in modelling data on tourist safety, combined with the use of geographical information systems (GIS), can assist in identifying the scale, magnitude and spatial distribution of accidents and risk.

Yet integrating all these perspectives within one specific place, such as this book, is too complex given the constraints of space. Specific concepts and issues from each discipline need to be incorporated where they can make a contribution to the conceptualisation, analysis and discussion of the factors that may be negatively or positively impacting upon the tourists' well-being. For this reason, this book is truly multidisciplinary in its approach to tourist health and safety and emphasises the interconnections between the tourist, tourism industry and the ways in which management and public sector interventions can ensure tourist well-being is safeguarded and enhanced.

Although the selection of chapters in this book is not necessarily illustrative of the scope, context and interventions in tourism-related activities to ensure that well-being can be enhanced, it does stimulate a wider debate within the area of tourism studies, and even wider in the public policy arena and throughout the social sciences which need to be integrated in the analysis of tourist well-being. Above all, the book intends to highlight the applied and rational role which research can play in hazard identification, the analysis of the scale, extent and nature of risks in tourism and the various interventions that might be appropriate to address such issues. We have assembled contributors from the wider area of tourism studies and what might loosely be called 'travel medicine' to provide a balance of practitioner and academic research inputs to the book. The editors, each of whom integrate both practical and applied research for industry and government agencies in this field with pure academic research, recognise the role which a blend of academic and practitioners can make to widening the area of travel medicine to the tourism field. This book will certainly bring this exciting, fast moving and rapidly expanding area to a wider audience.

The Structure of the Book

This book is structured as a series of informed perspectives on tourist health and safety that synthesise our current knowledge and provide a platform for management decisions. The four sections cover traditional areas such as Medical Conditions and Injuries (Section 1), as well as the emerging specialty area of Adventure Tourism (Section 2). Advice and Best Practice (Section 3) draws together a review of current issues in travel and tourism law, along with an analysis of the role of travel agents as an information source and some case studies of best practice in tourist safety. Selected Issues (Section 4) highlights broad topics of current interest such as transport safety and disaster management.

Chapter 1 has set the scene by outlining the four main sources of health, safety and security risks in tourism, as identified by the World Tourism Organization (2003). Of the four areas, physical safety and the prevention of injuries is a theme that runs throughout the text, as part of a larger interest in tourist well-being. The management of tourist health and safety in the new millennium will also rely heavily on active partnerships between government agencies, the tourism sector and local communities at travel destinations around the globe. As the recent World Tourism Organization (WTO) (2003) review confirms, governments will need to take the lead in this area, as tourism groups still struggle with admissions that their customers may become sick or injured in the course of their travels. Adequate information, preparation, delegation of tasks and responsibilities, and a willingness to tackle problem areas, including the development of an honest working relationship with the media, are the hallmarks of successful tourist health and safety management in the future.

References

Anonymous (2001). Sick leave. *Holiday Which?*, (Summer), 130–131.

Atherton, T. C., & Atherton, T. A. (1998). *Tourism, travel and hospitality law.* Sydney: LBC Information Services.

Clift, S., & Grabowski, P. (Eds) (1997). *Tourism and health: risks, research and responses.* London: Pinter.

Clift, S., & Page, S. J. (Eds) (1996). *Health and the international tourist.* London: Routledge.

Department of Industry, Tourism & Resources (2002). *The 10 year plan for tourism: A discussion paper.* Canberra: Department of Industry, Tourism & Resources.

DuPont, H. L., & Steffen, R. (Eds) (1997). *Textbook of travel medicine and health.* Hamilton, Ontario: BC Decker.

Federation of Tour Operators (1999). *Health and safety handbook.* Lewes, United Kingdom: Federation of Tour Operators.

Hall, C. M., & Page, S. J. (2002). *The geography of tourism and recreation: Environment, place and space* (2nd ed.). London: Routledge.

Hargarten, S. W., Baker, T. D., & Guptill, K. (1991). Overseas fatalities of united states citizen travellers: An analysis of deaths related to international travel. *Annals of Emergency Medicine, 20,* 622–626.

Hospedales, C. J. (1997). Destination: Healthy tourism in the Caribbean. *Perspectives in Health, 2* (2), 14–19.

International Hotel and Restaurant Association (1998). *Think-tank findings on safety and security. Executive summary.* Unpublished Report, Orlando, Florida, 18 & 19 August 1998.

International Hotel and Restaurant Association (2000). *Safety and security think tank discussion.* Unpublished Draft Report, Paris, 1 March 2000.

International Labour Organization (2001). ILO global employment forum faces up to shrinking job markets. www.ilo.org/public/english/bureau/inf/pr/2001/36.htm

Keystone, J., Kozarsky, P., Nothdurft, H. D., Freedman, D. O., & Connor, B. (Eds) (2003). *Travel medicine* (in press). New York: Mosby.

Leggat, P. A., & Leggat, F. W. (2002). Travel insurance claims made by travellers from Australia. *Journal of Travel Medicine, 9,* 59–65.

Liability Insurance Taskforce (2002). Report to the Queensland government. Available at www.premiers.qld.gov.au/about/pcd/economic/insurancetaskforce.pdf

MacPherson, D. W., Guérillot, F., Streiner, D. L., Ahmed, K., Gushulak, B. D., & Pardy, G. (2000). Death and dying abroad: The Canadian experience. *Journal of Travel Medicine, 7,* 227–233.

Nicol, J., Wilks, J., & Wood, M. (1996). Tourists as inpatients in Queensland regional hospitals. *Australian Health Review, 19,* 55–72.

Page, S. J., & Meyer, D. (1996). Tourist accidents: An exploratory analysis. *Annals of Tourism Research, 23,* 666–690.

Page, S. J., Bentley, T. A., Meyer, D., & Chalmer, D. J. (2001). Scoping the extent of tourist road safety: Motor vehicle transport accidents in New Zealand 1982–1996. *Current Issues in Tourism, 4* (6), 503–526.

Paixao, M. L., Dewar, R. D., Cossar, J. H., Covell, R. G., & Reid, D. (1991). What do scots die of when abroad? *Scottish Medical Journal, 36*, 114–116.

Pizam, A., & Mansfeld, Y. (1996). *Tourism, crime and international security issues*. Chichester, England: John Wiley & Sons.

Prociv, P. (1995). Deaths of Australian travellers overseas. *Medical Journal of Australia, 163*, 27–30.

Queensland Fire and Rescue Authority (2000). Building fire safety in Queensland budget accommodation: A report resulting from the fire in the Palace Backpackers Hostel, Childers. www.fire.qld.gov.au/about/pdf/QFRA-HO-Report Childers_Taskforce_Report.pdf

Ryan, C. (1996). Linkages between holiday travel risk and insurance claims: Evidence from New Zealand. *Tourism Management, 17* (8), 593–601.

Smith. D. M. (1977). *Human geography: A welfare approach*. London: Arnold.

Standards Australia and Standards New Zealand (1999). *Risk management. Australian/New Zealand Standard: AS/NZS 4360:1999*. Strathfield, New South Wales: Standards Association of Australia.

Surf Life Saving Queensland (2001). *The life of the beach: Annual report 2000–2001*. Brisbane: Surf Life Saving Queensland.

Taylor, H. (2001). Many people unprepared for terrorist attacks or other disasters. *The Harris Poll, 60* (5 December). Available at www.harrisinteractive.com/harris_poll/searchpoll.asp

Trowbridge Consulting (2002). Public liability insurance. Analysis for meeting of ministers 27 March 2002. Available at www.trowbridge.com.au/dir095/home.nsf

Wanderlust Magazine (2001). September 11 — The aftermath. Available at www.wanderlust-.co.uk/survey/sept11.html

Wilks, J. (1999). International tourists, motor vehicles and road safety: A review of the literature leading up to the Sydney 2000 Olympics. *Journal of Travel Medicine, 6*, 115–121.

Wilks, J. (2003). Injuries and injury prevention. In: J. Keystone, P. Kozarsky, H. D. Nothdurft, D. O. Freedman, & B. Connor (Eds), *Travel Medicine* (in press). New York: Mosby.

Wilks, J., & Coory, M. (2000). Overseas visitors admitted to Queensland hospitals for water-related injuries. *Medical Journal of Australia, 173*, 244–246.

Wilks, J., & Coory, M. (2002). Overseas visitor injuries in Queensland hospitals: 1996–2000. *Journal of Tourism Studies, 13*, 2–8.

Wilks, J., & Davis, R. J. (2000). Risk management for scuba diving operators on Australia's Great Barrier Reef. *Tourism Management, 21*, 591–599.

Wilks, J., & Oldenburg, B. (1995). Tourist health: The silent factor in customer service. *Australian Journal of Hospitality Management, 2*, 13–23.

Wilks, J., Pendergast, D. L., & Wood, M. T. (2002). Overseas visitor deaths in Australia: 1997–2000. *Current Issues in Tourism, 5*, 550–557.

Wilks, J., Walker, S., Wood, M., Nicol, J., & Oldenburg, B. (1995). Tourist health services at tropical island resorts. *Australian Health Review, 18*, 45–62.

World Health Organization (1978). *International classification of diseases* (9th rev.). Geneva: World Health Organization.

World Tourism Organization (1997). *Tourist safety and security: Practical measures for destinations* (2nd ed.). Madrid: World Tourism Organization.

World Tourism Organization (2002). Facts and figures. www.world-tourism.org/market_research/facts&figures/menu.htm

World Tourism Organization (2003). *Safety and security in tourism: Partnerships and practical guidelines for destinations*. Madrid: World Tourism Organization, in press.

World Tourism Organization and World Meteorological Organization (1998). *Handbook on natural disaster reduction in tourist areas*. Madrid: World Tourism Organization.

Section 1

Medical conditions and injuries

Chapter 2

Travellers' Health: An Epidemiological Overview

Johnathan Cossar

Introduction

While there have been notable medical advances throughout the twentieth century, the contemporary traveller is still vulnerable to health hazards resulting from the very nature of travel itself. Travel exposes the individual to new cultural, psychological, physical, physiological, emotional, environmental and microbiological experiences and challenges. The travellers' ability to adapt, cope and survive these challenges is influenced by many variables, including their pre-existing physical, mental, immunological and medical status. This is subsequently modified by personality, experience and behaviour and differs according to age, gender, culture, race, social status and education. The final aspect of this challenge relates specifically to the unfamiliar environmental exposure which encompasses climate, altitude, sunlight, hygiene, and disease prevalence. For these reasons, it is not surprising that health problems affecting the traveller have been recognised throughout history.

Historical Aspects

There are many historical examples of health hazards affecting travellers and, indeed, influencing history. For example, in the twelfth and thirteenth centuries the authorities in Venice noted that there was a problem with regular outbreaks of plague affecting the city's inhabitants, which occurred shortly after the arrival of ships from the East. Consequently, Venice and Rhodes introduced the first regulations governing the arrival of ships in 1377. Ships were detained at a distance, complete with passengers, cargo and crew for forty days (*quaranto giorni*), before gaining permission to proceed to their final destination (Bruce-Chwatt 1973). This measure is generally accepted as the conceptual origin of quarantine and other cities and countries followed this example until some

Managing Tourist Health and Safety in the New Millennium
Copyright © 2003 by Elsevier Science Ltd.
All rights of reproduction in any form reserved.
ISBN: 0-08-044000-2

form of sanitary regulation became commonplace in many countries during the next five centuries.

Further travel hazards were also noted in later times. The ill-fated Darien Expedition of the 1690s was a disastrous attempt to establish a Scottish colony on the Isthmus of Panama. Some 2,000 Scots died as a result of appalling local conditions where malaria and yellow fever were rife (Steel 1984).

At the turn of the last century, considerable numbers of citizens from the United Kingdom were engaged in work abroad. Africa presented notable health risks to them; for example, the missionaries Mungo Park, David Livingstone, and Mary Slessor all died in Africa, succumbing respectively to trauma (drowned while under attack by hostile natives), dysentery with internal haemorrhage and 'exhaustion'.

More detailed study of 1,427 Scottish Presbyterian missionaries working abroad between 1873 and 1929 revealed that 25% returned prematurely due to personal or family ill health and a further 11% died in service (Cossar 1987). In addition, the numbers affected by adverse health were greater for those appointed in the earlier years, when less was known about tropical diseases and the problems were more severe for those appointed to the most climatically rigorous areas; for example, Western Africa. Interestingly, missionaries with a medical background experienced fewer problems, probably because of their knowledge of illness and its prevention. The effect of local climate and environment, and background knowledge about disease, continue to be very relevant to the health experience of the contemporary traveller.

The Growth of Travel

In 1949, 26 million international tourists were recorded (Table 2.1) whereas by 2000 this had risen to 699 million (World Tourism Organization 2000) representing a 27 fold increase, with 58% of tourists travelling to the European region. During this time there was also a 53 fold increase in scheduled air travellers (International Civil Aviation Organisation 2000) and a 33 fold rise in United Kingdom residents travelling abroad, with the proportion of those travelling beyond Europe increasing 119 fold (Business Statistics Office 2000). Those exposed to the wider extremes of climate and culture are showing the most accelerated growth rate. Groups contributing to this growth in travel include tourists, political representatives, business people, technical experts, pilgrims, migrant workers, students, refugees, immigrants, military personnel, sporting partici-pants and followers, and the hotel and travel support services.

Table 2.1: Growth in international travel 1949–2000 (in millions).

	1949	1960	1970	1980	1990	2000
Global air travellers	31.0	106.0	386.0	748.0	1160.0	1647.0
International tourists	26.0	69.0	160.0	285.0	429.0	699.0
Visits abroad by United Kingdom residents	1.7	6.0	11.8	17.5	31.2	56.8

The increasing speed, capacity and frequency of modern travel, coupled with the increased numbers of travellers makes it more likely than previously that travellers will return within the incubation period of many infections. For example, between 1948 and the present, the fastest passenger-aircraft cruising speed has risen from 340 to 1,356 miles per hour (supersonic travel), and the maximum capacity from 40 to over 400 passengers (jumbo jet). The long return sea journey enabling illnesses acquired abroad to be recognised before return is now relatively rare and travellers and doctors need to be more alert to this possibility.

Health Implications of Travel

This contemporary travel phenomenon not only has economic, environmental, cultural and social repercussions but also medical, epidemiological and medico-legal consequences. Some illnesses may be induced by travel itself, such as motion sickness and upsets to the circadian rhythms; unaccustomed exercise or the effects of altitude may exacerbate pre-existing cardiovascular or respiratory pathology.

The effects of exposure to unfamiliar infectious agents and the stress of altered climate, environment and culture may also cause problems for the unwary traveller, which may be compounded by differing medical practices encountered internationally. In addition, the media are always interested to dramatically report on imported infections, such as the 2,000 + malaria notifications and seven deaths recorded annually in the United Kingdom (PHLS 1997) of which 68% result from *P.falciparum* or the 81% of heterosexually Acquired Immune Deficiency Syndrome (AIDS) cases (3,700 total cases to 30 September 2001) (SCIEH 2002) which were contracted abroad (the same pattern as seen for HIV infection). However, it is only through epidemiological studies that the true perspective of diseases associated with travel can be defined.

Epidemiological Studies of Travellers

Travellers who return with a previously unknown disease can present diagnostic difficulties and life-threatening delays in the home country. For example, in 1973, a group of package holidaymakers returning from Benidorm, Spain to Glasgow, Scotland suffered an outbreak of pneumonia and three fatalities resulted. These deaths were subsequently attributed to Legionnaires disease (Lawson *et al.* 1977). The deaths motivated the Communicable Diseases (Scotland) Unit (CDSU), the University of Glasgow's Department of Infectious Diseases, and the Department of Laboratory Medicine and the Regional Virus Laboratory, both at Ruchill Hospital, Glasgow to collaboratively study illnesses associated with travel. Over the past 25 years the CDSU, now the Scottish Centre for Infection and Environmental Health (SCIEH), has established systems to monitor the health experience of returning Scottish travellers and make specific enquiries into groups of travellers identified as being 'at risk' following an alert about a possible health problem (Cossar 1989; Cossar & Reid 1989; Cossar *et al.* 1982; 1984; 1988a; 1998; 1990; Dewar *et al.* 1983; Grist *et al.* 1985). This epidemiological overview is based on the findings from this and other research programmes.

A series of studies of 14,227 Scottish holidaymakers showed that the overall attack rate was 37%. Of these, 28% suffered from alimentary symptoms, 3% from respiratory illness, and 5% had both (Cossar *et al.* 1990) (Figure 2.1). In a Finnish investigation of 2,665 travellers, 18% reported illness (Peltola *et al.* 1983).

When age was considered in the Scottish studies, the highest attack rates of illness were recorded by those aged less than 40 years, with 41% of the 10 to 19 years age groups, 48% of the 20–29 years age groups, and 38% of the 30 to 39 years age groups reporting illness (Table 2.2). Thereafter attack rates showed a progressive diminution with increasing age. In short-term Swiss travellers to developing countries, the highest rate (20%) was also in the 20 to 29 years age group (Steffen *et al.* 1983).

When the country of destination was considered, there was a general indication that the further south one travelled and, to some extent, the further east in Europe and beyond, the higher the rate of illness (Table 2.3) and this generally remained true in both summer and winter. Examples in support of this trend were the 77% attack rate reported by tourists to Northern Africa in the summer and the 57% rate for those travelling to Eastern Europe. In Swiss travellers, the highest rate of illness (21%) was noted in those returning from East Africa, Sri Lanka, and the Far East, and almost three times that of travellers returning from the Greek Islands (Steffen *et al.* 1987). Reports of illness are less frequent during the winter months (Cossar *et al.* 1998a).

The broad correlation of these various findings lends credibility to attempts at detecting patterns of illness derived from the travellers studied, and the use of largely

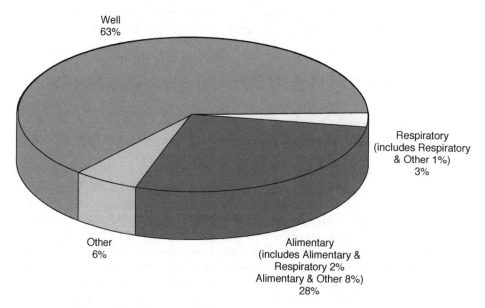

Figure 2.1: Reports of illness on Scottish tourists 1973–1985.

Table 2.2: Age of travellers and reports of illness among Scottish tourists 1973–1985.

Age group	Total respondents	% unwell
0–9	550	33
10–19	1974	41
20–29	3033	48
30–39	2028	38
40–49	2297	32
50–59	2381	28
60+	1239	20
not known	725	32
Total	14227	37

identical methodologies encourages a comparative analysis of relative risk of illness to the travellers, as shown in Table 2.4.

Mortality in Travellers

Travel related deaths can occur on the way to the destination, within the country being visited, or soon after return. In colonial times, mortality was very high among overseas travellers of whom the majority worked for prolonged periods as missionaries, military

Table 2.3: Area visited, season, and reports of illness among Scottish tourists 1973–1985.

Area visited	Summer attack rate %	Winter attack rate %
Europe (North)	19	20
Europe (East)	57	12
Mediterranean (Southern Europe)	34	19
Mediterranean (North Africa)	77	32
Average attack rates	77	20

Table 2.4: Profile of travellers at risk emerging from Scottish research.

Package holidaymakers versus other travellers
Inexperienced travellers versus other travellers
Travellers further south, particularly Northern Africa, versus other travellers
Summer travellers versus winter travellers
Younger age groups (specifically 20–29 years) versus older travellers
Smokers versus non-smokers

(Clift & Page 1996: 32)

personnel, and colonial administrators; short-stay overseas vacations were the exception. Death in these early travellers was often due to infection.

Although a large number of travellers become infected abroad nowadays, it is paradoxical that there are so few infection-related deaths. In a study of 421 deaths among Australian travellers, only 10 (2.4%) were attributed to infection (Prociv 1995). Similarly, in a Scottish study involving 952 persons who died abroad, infection occurred in 34 (3.6%) (Paixao *et al.* 1991). Men die more frequently than women do while abroad — almost 4 : 1 and 3 : 1 in the Australian and Scottish studies, respectively.

By far the most common cause of death was cardiovascular disease, usually myocardial infarction (35% in the Australian study and 68.9% in the Scottish). Among American travellers, cardiovascular problems among male travellers aged over 60 years accounted for 50% of reported deaths (Jong & McMullen 1995).

Trauma has emerged as an important cause of death in travellers, and various studies have pointed to the fact that motor vehicle accidents and other injuries cause significant mortality (Carey & Aitken 1996; Gangarosa *et al.* 1980; Guptill *et al.* 1991; Hargarten *et al.* 1991; Jong & Mc Mullen 1995; Odero *et al.* 1997; Paixao *et al.* 1991; Prociv 1995; Sniezek & Smith 1991). In American travellers, motor vehicle accidents accounted for 25% of all deaths, with other injuries and accidents such as drowning and fall from a height causing 10% (Jong & McMullen 1995). Among Australian travellers, accidents (mainly traffic accidents) accounted for 18% of deaths, and in Scottish travellers 21%. When those with cardiovascular disease are compared with those who had injuries or accidents, there is a marked preponderance of older persons in the former and younger persons in the latter.

It is not surprising that most injuries and accidents occur in younger age groups who are often involved in more active pursuits. Traumatic deaths such as road traffic accidents do seem to be a major hazard for younger travellers, and more attention should be drawn to ways in which they can avoid hazardous situations. Although it is possible that those who succumb to cardiovascular illness would have died anyway had they stayed at home, perhaps more consideration should be given to the question of whether those with pre-existing cardiovascular problems are wise to seek very warm climates, which may present an added workload to an already stressed heart. Since, in practice, no preventive measures guarantee safety, adequate health insurance should be obtained before venturing abroad.

Table 2.5: Travel associated admissions to infectious disease wards Glasgow 1985; 1998–1999.

	1985		1998/99	
Total admissions:	1249		1316	
Travel associated admissions	55	4%	108	8%
Details of travel associated admissions:		%		%
Sex:				
Male	34	62%	80	74%
Female	21	38%	28	26%
Ethnic origin:				
African/Carribean	4	7%	2	2%
Asian/Oriental	30	55%	19	18%
Caucasian	21	38%	87	81%

Hospital In-Patient Data

Studies of data from patients admitted to hospital with travel related illnesses provide another perspective (Table 2.5). In-patient data from the major infectious diseases facility for the Greater Glasgow area were analysed in 1985 and again in 1998–1999 (Cossar *et al.* 2002). The hospital's catchment area covers approximately 15% of the Scottish population and 750,000 people.

During the period 1985 to 1998–1999, United Kingdom travel statistics showed a 135% growth in visits abroad and a 5% rise to 17% in destinations with a risk of malaria. While the overall infectious disease in-patient total rose by 5%, travel associated admissions rose by 96% to total 108. Patients of Asian/Oriental ethnicity declined from 55% to 18%, while Caucasians increased from 38% to 81%. Travellers aged 20–29 years had the highest attack rates, 51% and 50%. Gastro-intestinal problems accounted for the largest single diagnostic category in each study period, 38% and 40% respectively. In-patients diagnosed with malaria fell by 20%, which is encouraging to those involved in the teaching and dissemination of advice on malaria prophylaxis. These findings highlight the need for the continuation and expansion of travel health education for both healthcare professionals and the public.

An average cost per travel related in-patient was calculated for both time spans (£428 and £2,136 respectively). These figures were then extrapolated to calculate a theoretical total for United Kingdom hospitalisation costs from travel related admissions. This totalled £81.95 million in 1998/9 compared with £15.3 million in 1985. However, there are clear limitations in this generalisation to the whole United Kingdom as these figures relate to inclusive tour holiday/package travellers only, the study numbers are small, and no account is made for potential demographic differences.

Protecting Travellers

While it is important travellers are advised about the vaccinations appropriate to their destination, it should be emphasised that immunisation and medication can, at best, only protect against about 5% of the health hazards to which travellers are exposed. Therefore the potential opportunities for pre-travel health education are vast.

Preparation for Travel

Good insurance cover is strongly recommended for all who travel and it is a false economy to ignore the issue as the costs of medical care abroad can be financially crippling. Adequate medical insurance is required as the standard of local medical care can be highly variable and it is in the best interests of the traveller to be speedily repatriated.

In terms of specifying risk, it is worth noting that insurance statistics from the United States of America suggest that three in twenty persons, irrespective of travel, can expect to develop an illness of minor or major significance, every two weeks. This is of greater significance in travellers with a pre-existing illness who are more vulnerable to develop a health problem, such as the young, the elderly and the pregnant traveller.

It is important that intending travellers consult for medical advice regarding immunisations, at least 4–6 weeks in advance to ensure that there is adequate time to complete any complex immunisation schedules. A recent dental check-up will minimise the risk of a painful, and possibly expensive experience abroad, as well as anxiety about the safety of needle and instrument sterilisation.

The Journey

The enforced immobility of a long plane journey encourages pooling of fluid in the dependent legs, which in turn predisposes to the development of ankle swelling and deep venous thrombosis. Travellers with venous insufficiency problems, such as the elderly and obese are more at risk, and it is recommended that they walk about the aircraft for about five minutes during every hour of travel as well as wearing appropriate support hose. Additionally, those on daily medications should carry adequate supplies in their hand luggage; otherwise lost baggage can lead to medical complications in addition to frustration.

The modern passenger jet is pressurised to the equivalent of an altitude of 5,000–7,000 feet. This rarely presents a problem for the fit traveller, but the reduced oxygen saturation may present difficulties for those with impaired oxygen carrying capacity (e.g. severe bronchitis) or requiring good oxygen saturation (e.g. recent heart attack) which is aggravated in the presence of cigarette smoking. Due to the system of air circulation within the aircraft, dehydration can be a problem unless adequate fluids are regularly consumed during a long journey. Finally, an unexpected stopover may occur, exposing the unprepared traveller to disease risk such as malaria.

Safe Water

Water, being a pre-requisite for life, can also present particular health hazards for the traveller. Direct ingestion of contaminated water as well as indirect from food consumption of, for example fish, are the most obvious sources of problems. Diseases caused by such ingestion include diarrhoeal diseases, typhoid, cholera, poliomyelitis, Hepatitis A and worm infections.

Some simple avoidance strategies include drinking only 'brand name' carbonated bottled waters; boiling, which kills all infective agents including amoebic cysts; use of commercially available filters, and chemical disinfections (e.g. 4 drops of 2% iodine to 1 litre of water and allow to stand for 20–30 minutes). Such water should also be used for teeth brushing.

Similarly, locally obtained fresh food should be regarded as contaminated. Fruit and vegetables are best consumed after thorough washing in treated water, or (for salad foods), soaking (12 drops of 2% tincture of iodine to 1 of litre water and allowed to stand), before peeling. Raw vegetables and cold food prepared by others are best avoided. Seafood, fish and meat should be consumed well cooked, and unpasteurised milk avoided or boiled.

There are also problems caused by the lack of adequate hygiene and sanitary facilities, which predispose to skin and other infections. An additional water-related concern is bites or penetration by water-breeding insects that cause disease, e.g. malaria, schistosomiasis.

Sea

Apart from the obvious hazards from recreational pursuits in unsuitable maritime environments there can now be additional risks as a consequence of modern industrial/ agricultural practices. Specifically, chemical contamination of recreational water can give rise to illness such as gastro-enteric upset, although the risk from ingestion of affected seafood is the more likely source of trouble. Also, the intensive use of nitrogen-based fertilisers means that the cumulative effect from this being washed off the land predisposes to the development of algal blooms. These have the potential to have an irritant effect on skin and respiratory tissues with the added complication of secondary infection. For these reasons prolonged exposure to contaminated air from such areas or bathing is inadvisable. Any clothing exposed to algae from water sports should be thoroughly hosed down after use. Clearly in areas where there is venomous sea life appropriate precautions should be taken.

Sun

Awareness of the sun's ability to cause not only short term discomfort from burning but also long term risks should be fully appreciated. Excessive exposure predisposes to premature skin ageing, skin cancer and cataract development. It is now realised that even a single episode of sun burning in a child carries a disproportionate risk of future skin cancer compared with an adult. Sunlight is also synergistic to fungal skin infections. Appreciation of these facts by those with high photosensitivity, as well as the amplification of the power of sunlight by the effects of water, snow, sand, altitude, latitude and medication will promote a sensible approach in terms of protective clothing, gradual exposure and the use of appropriate sunscreens.

Sex

On account of the sensitive personal nature of this topic, it is one which many advice sources prefer to avoid. However, it has long been recognised that travel is directly associated with making new sexual contact(s), and therefore by definition, is in itself a risk factor. Once this is accepted then some of the sensitivities can be put aside. At the same time it seems unfair not to inform the traveller that most of the countries likely to be visited have a higher prevalence rate of HIV than the United Kingdom; that over 80% of the cumulative total of heterosexually contracted AIDS reported in the United Kingdom to date were acquired abroad; and that women are up to four times more susceptible than men of contracting AIDS from unprotected intercourse.

Thereafter specific risk enhancing factors, such as the greater the number of partners the greater the risk, the higher the frequency of sexual contact the higher the risk, and the choice of a high risk partner (a prostitute, drug abuser, male homosexual/bisexual), can be more fully appreciated. The messages to be propagated are abstinence, sexual intercourse with one, faithful, lifetime partner, or always carry and use condoms. However, this is an area where much remains to be done in appropriate targeting with suitable material before there can be any realistic hope of modifying travellers' sexual behaviour.

Malaria

The important message about malaria is that no current anti-malarial measures can guarantee absolute protection, although if all mosquito bites can be avoided, then infection will not occur. Even newer anti-malarials have limitations, as well as recommendations (Bradley 1993). Therefore, as well as taking the appropriate anti-malarial for 1–3 weeks prior to entering a malarious area and continuing until 4 weeks following departure from same, the traveller should take every precaution to minimise mosquito bites.

These precautions include covering exposed skin from dusk onwards and applying an insect repellent containing diethyl toluamide (DEET) both to the skin and clothing. The use of impregnated wrist and ankle bands is additional protection for vulnerable areas. This is then complemented by the use of permethrin impregnated mosquito nets over the bed, employing a knock-down spray (pyrethroids) to the room prior to retiring, and a vapona/burning a coil during sleep which is effective against mosquitoes in sealed rooms.

A further more recent concept is the use of stand-by medication provided by the doctor prior to travel for those at risk from resistant malaria in remote locations where rapid access to diagnostic facilities is not possible. The medication is commenced at the onset of typical symptomatology (temperature, fever, chills) while locating the nearest medical facility for further help. Any traveller returning from a malarious area presenting with a flu-like illness should be regarded as suffering from malaria until proved otherwise.

Accidents

Accidents are the second most common cause of death overall, as well as the premier cause of death in younger travellers. Paixao *et al.* (1991) reviewed 952 deaths in travellers and reported that 21% of the deaths were caused by trauma, and within this group 78 (40%) were accounted for by the 10–29 year age group. With better insight through pre-travel health education it should be possible to reduce this tragic annual toll of avoidable deaths.

Simple raised awareness of the consequences of a moments inattention/carelessness, such as diving into unfamiliar swimming pools, not wearing seat belts/crash helmets, failure to remember the contra-traffic direction to the home country, appreciation of differing environmental safety standards (e.g. balcony barrier heights), bites from poorly domesticated animals (perhaps rabid) and exposure to risk of assault by straying from the recognised tourist areas, can minimise preventable accidents. Deaths from assault received a high press profile in 1993, with nine such tourist deaths in Florida in nine months. Much can be done by travel agents/tour operators/travel couriers to alert the unwary tourist to unsafe/no go areas and to injudicious conduct.

All of the above hazards are potentiated by the injudicious use of alcohol, and the natural adventurousness, energy, and disinhibition of youth clearly highlight the prime at risk group.

Sources of Information and Advice

An inevitable accompaniment to the growth in international travel is that general practitioners and other primary care workers are increasingly contacted by patients seeking advice both prior to travel and following their return, and the intending traveller is faced with a plethora of sources of advice (Figure 2.2).

Although the travel agent may be the most commonly consulted advice source (Cossar *et al.* 1990), there is concern about the availability and quality of that advice (Reid *et al.* 1986). It is clear that no one source can address all the diverse needs of the traveller, rather a case of horses for courses, but that the general practitioner occupies a pivotal role in this field. There are studies that suggest that general practitioners may encounter problems about giving appropriate advice (Usherwood & Usherwood 1989); that general practice may not be the best location for provision of travel advice (Jeffries 1989), but that general practitioners have a medico-legal responsibility to provide accurate advice (Holden 1989).

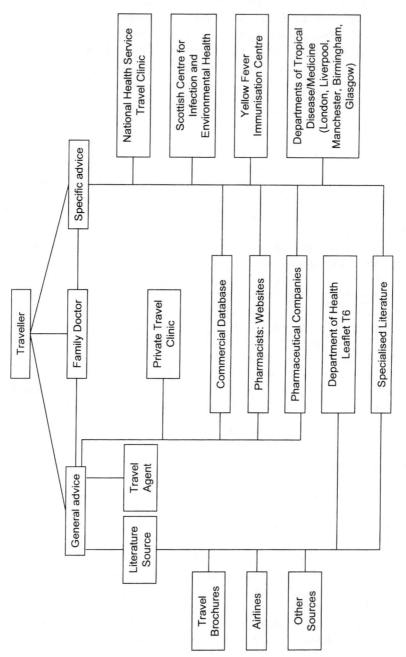

Figure 2.2: Sources of pre-travel health advice for travellers.

In recognition of the responsibilities and difficulties which the general practitioner faces in addressing this need, the CDSU commenced a telephone advice service for the primary care sector in 1975. In addition, a computerised health information database (Travax) (Cossar *et al.* 1988b) was established in 1982. This nationally accessible database includes recommendations on vaccinations for individual countries, malaria prevention advice and information about particular vaccines to help in making a balanced judgement when indications are not clear-cut. Details about administration and schedules, interactions between vaccines, and advice for those with HIV infection are included, along with current notes detailing known outbreaks, and changes in vaccine availability. This service (Travax) is available to both National Health Service personnel and commercial users.

A study by Arnold (1990) involving 899 travellers departing the United Kingdom from Heathrow airport showed a 65% preference for pre-travel health advice to be provided by the general practitioner. This is logical, as the general practitioner is uniquely placed to advise, having access to the relevant past medical history, including previous immunisations, allergic reactions, long-term medication and past illnesses, as well as information on the patient/travellers lifestyle. Such information is essential for the provision of accurate, appropriate advice for the individual traveller.

At the same time it has to be recognised that it is impractical for all travellers to attend the family doctor prior to departure and indeed there is much information to be disseminated apart from immunisation advice. Professionals other than the general practitioner may even better present such advice. It is therefore important that every opportunity for the dissemination of appropriate pre-travel health advice be taken whenever the consumer interfaces with the travel scene, be this the travel agent, airline, currency supplier, visa/government agent, or specifically dedicated medical facilities. Another recent innovation is a web-based service for the general public, Fit for Travel, based on the Travax database, which was launched in 2000 (www.fitfortravel.scot.nhs.unitedkingdom).

Concluding Observations

The growth of travel and the numbers affected by travel-related illnesses, some of a serious nature, means that this subject will increasingly demand the recognition of the medical profession, health promotion specialists, the travel trade and travellers. Provision of appropriate advice for the traveller, which is already available, is a shared responsibility and is best channelled through all the agencies which interact with travellers.

Continued monitoring of illness in travellers and provision of information systems about this problem and its prevention, utilising traditional communication channels as well as modern computer technology, with ready access for medical and related workers, is fully justified. Increased collaboration on travel illness between medical workers, health educators and those involved in the travel trade, will be a very positive and efficacious contribution to reducing illness in, and discomfort for, travellers, and the associated expense that this brings to the health services.

References

Arnold, W. S. J. (1990). *Vaccine information and the practice nurse.* Paper presented at the Third International Conference on Tourist Health, Venice.

Bradley, D. (1993). Prophylaxis against malaria for travellers from the United Kingdom. *British Medical Journal, 30* (6), 1247–1252.

Bruce-Chwatt, L. (1973). Global problems of imported disease. *Advances in Parasitology, 11,* 75–114.

Business Statistics Office (2000). *Business monitor annual statistics 17: Overseas travel and tourism (MQ 6).* London: Business Statistics Office.

Carey, M. J., & Aitken, M. E. (1996). Motorbike injuries in Bermuda. *Annals of Emergency Medicine, 28,* 424–429.

Cossar, J. H. (1987). *Studies on illnesses associated with travel.* Unpublished doctoral dissertation, University of Glasgow, Glasgow.

Cossar, J. H. (1989). A review of travel associated illness. In: *Travel medicine. Proceedings of the first conference on international travel medicine, Zurich* (pp. 50–54). Berlin: Springer-Verlag.

Cossar, J. H., & Reid, D. (1989). Health hazards of international travel. *World Health Statistics Quarterly, 42* (2), 27–42.

Cossar, J. H., Dewar, R. D., Fallon, R. J., Grist, N. R., & Reid, D. (1982). Legionella pneumophila in tourists. *Practitioner, 226,* 1543–1548.

Cossar, J. H., Dewar, R. D., Fallon, R. J., Reid, D., Bell, E. J., Riding, M. H., & Grist, N. R. (1984). Rapid response health surveillance of Scottish tourists. *Travel and Traffic Medicine International, 21,* 23–27.

Cossar, J. H., Dewar, R. D., Reid, D., & Grist, N. R. (1988a). Travel and health: Illness associated with winter package holidays. *Journal of the Royal College of General Practitioners, 33,* 642–645.

Cossar, J. H., Reid, D., Fallon, R. J., Bell, E. J., Riding, M. H., Follett, E. A. C., Dow, B. C., Mitchell, S., & Grist, N. R. (1990). A cumulative review of studies on travellers, their experience of illness and the implications of these findings. *Journal of Infection, 21,* 27–42.

Cossar, J. H., Walker, E., Reid, D., & Dewar, R. D. (1988b). Computerised advice on malaria prevention and immunisation. *British Medical Journal, 296,* 358.

Cossar, J. H., Wilson, E., Kennedy, D. H., & Walker, E. (2002). A comparison of travel related admissions in Glasgow: (1985; 1998/99). *The Scottish Medical Journal,* (December).

Dewar, R. D., Cossar, J. H., Reid, D., & Grist, N. R. (1983). Illness amongst travellers to Scotland: A pilot study. *Health Bulletin* (Edinburgh), *41* (3), 155–162.

Gangarosa, E. J., Kendrick, M. A., Loewenstein, M. S., & Mosely, J. W. (1980). Global travel and travellers' health. *Aviation, Space and Environmental Medicine, 51,* 265–270.

Grist, N. R., Cossar, J. H., Reid, D., Dewar, R. D., Fallon, R. J., Riding, M. H., & Bell, E. J. (1985). Illness associated with a package holiday in Romania. *Scottish Medical Journal, 30,* 156–160.

Guptill, K. S., Hargarten, S. W., & Baker, T. D. (1991). American travel deaths in Mexico: Causes and prevention strategies. *Western Journal of Medicine, 154,* 169–171.

Hargarten, S. W., Baker, T. D., & Guptill, K. S. (1991). Overseas fatalities of United States citizen travellers: An analysis of deaths related to international travel. *Annals of Emergency Medicine, 20,* 622–626.

Holden, J. D. (1989). General practitioners and vaccination for foreign travel. *Journal of the Medical Defence Union,* (Spring), 6–7.

International Civil Aviation Organization (2000). *Development of world schedule revenue traffic, 1945–2000 statistics.* Montreal: ICAO.

Jeffries, M. (1989). Booster for GP travel vaccine clinics. *Monitor, 2* (31), 10–11.

Jong, E. C., & McMullen, R. (1995). *The travel and tropical medicine manual.* Philadelphia: W. B. Saunders.

Lawson, J. H., Grist, N. R., Reid, D., & Wilson, T. S. (1977). Legionnaires disease. *Lancet, ii,* 108.

Odero, W., Garner, P., & Zwi, A. (1997). Road traffic injuries in developing countries: A comprehensive review of epidemiologic studies. *Tropical Medicine and International Health, 2,* 445–460.

Paixao, L., Dewar, R. D., Cossar, J. H., Covell, R. G., & Reid, D. (1991). What do Scots die of when abroad? *Scottish Medical Journal, 36,* 114–116.

Peltola, H., Kyronseppa, H., & Holsa, P. (1983). Trips to the south — A health hazard. *Scandinavian Journal of Infectious Diseases, 15,* 375–381.

Prociv, P. (1995). Deaths of Australian travellers overseas. *Medical Journal of Australia, 163,* 27–30.

Public Health Laboratory Service (PHLS) Communicable Disease Surveillance Centre (1997). *Communicable Diseases Review, 7/10,* R139.

Reid, D., Cossar, J. H., Ako, T. I., & Dewar, R. D. (1986). Do travel brochures give adequate advice on avoiding illness? *British Medical Journal, 293,* 1472.

Scottish Centre for Infection and Environmental Health (2002). *SCIEH weekly report surveillance report HIV infection and AIDS: Quarterly report to 31 December 2001, 36* (2002/02), 13.

Sniezek, J. E., & Smith, S. M. (1991). Injury mortality among non-U.S. residents in the United States 1979–1984. *International Journal of Epidemiology, 19,* 225–229.

Steel, T. (1984). *Scotlands story.* London: Collins.

Steffen, R., Rickenbach, M., Willhelm, U., Helminger, A., & Schar, M. (1987). Health problems after travel to developing countries. *Journal of Infectious Diseases, 156,* 84–91.

Steffen, R., van der Linde, F., Syr, K., & Schar, M. (1983). Epidemiology of diarrhoea in travellers. *Journal of the American Medical Association, 249,* 1176–1180.

Usherwood, V., & Usherwood, T. P. (1989). Survey of general practitioners advice for travellers to Turkey. *Journal of the Royal College of General Practitioners, 39,* 148–150.

World Tourism Organization (2000). *Tables.* Madrid: World Health Organization.

Chapter 3

Evaluating the Nature, Scope and Extent of Tourist Accidents: The New Zealand Experience

Stephen J. Page, Tim Bentley and Denny Meyer

Introduction

The welfare of tourists has emerged as a growing area of interest for researchers from a wide range of disciplines including tourism, ergonomics, safety science, medicine and health studies as reflected in the expanding literature in books and journals. Yet within the wider concern for tourist welfare there has been a dearth of published literature concerning tourist accidents. This is surprising in light of growing evidence that adverse experiences (such as accidents) can have a significant negative effect on the image of the tourism industry (e.g. Wilks *et al*. 1996). This chapter briefly reviews some of the principal studies published on tourist accidents and safety, the way in which tourist accidents can be conceptualised. The chapter then reviews the scope and extent of tourist accidents within one country — New Zealand — using a range of disparate data sources which identify different dimensions of the tourist accident problem.

The Tourist Accident Literature

Paixao *et al*. (1991) reported that of the 952 travellers from Scotland who died abroad between 1973 and 1988, 21% of fatalities were due to accidents and injuries. This represented the second largest cause of death; the most common cause being cardiovascular disorders (69%). Road traffic accidents and violence (32%) were the main accident causes identified, with highest proportions of accidents incurred by the 20–29 age group. Males accounted for 80% of accident cases. A larger scale study undertaken by Hargarten, Baker & Guptill (1991) examined the causes of 2,463 deaths of American tourists while travelling overseas between 1975 and 1984. The study concurred with the findings by Paixao *et al*. (1991) that cardiovascular incidents were the major cause of death, while accidents comprised 25% of fatalities. Road traffic

Managing Tourist Health and Safety in the New Millennium
Copyright © 2003 by Elsevier Science Ltd.
All rights of reproduction in any form reserved.
ISBN: 0-08-044000-2

accidents and drownings were the most frequent cause of accidental death, with male travellers again accounting for the greater proportion of fatalities due to accidents (70%). Travellers over 65 years of age constituted 65% of all deaths, while the highest rates for injury deaths were found for the 15–24 years and 35–44 years age groups. Injury rates for travellers were significantly higher than United States of America death rates due to injury, while cardiovascular rates were significantly lower, suggesting an increased risk of injury-related deaths for overseas visitors.

A more recent study by Nicol *et al.* (1996) which analysed medical record data for seven Queensland (Australia) regional hospitals found the main reasons for overseas visitors' admissions were injuries and poisoning (37.6%). Accident injury was also the second most common reason for admission for interstate tourists (15.4%). The most common injury types were fractures (41%), decompression illness (13.4%), and drownings or non-fatal submersions were more common among overseas visitors (3.5%) than interstate tourists (1.3%).

Little published research has considered the accident and injury risk associated with specific recreational tourism activities with a number of notable exceptions (Bentley & Page 2001; Bentley *et al.* 2001). Recent research (Page & Meyer 1996; 1997) has considered the problem of tourism accidents in New Zealand, based primarily on analysis of data from the Accident Rehabilitation and Compensation Insurance Scheme (ACC). Recreational and sporting activities were identified as the most common type of claim for overseas visitors (41% of claims by overseas visitors), while New Zealand residents ('non-work' injuries) claimed most frequently for home accidents. A worrying feature is that visitors from the lucrative Japanese and American tourism markets registered the highest claim rates. Yet accidents are far from a simplistic notion and some attempt to conceptualise tourist accidents is necessary prior to any analysis of the issue so that the parameters and likely causes/explanations of accidents can be set in context.

Conceptualisation of Tourist Accidents

The definition and scope of accidents in tourism was provided by Page & Meyer (1996) and is reiterated here to show the multidisciplinary scope of the literature one has to consult and how a synthesis of these different perspectives can assist in deriving a holistic understanding of tourist accidents. An accident can be defined as 'anything that happens without foresight or expectation; an unusual event which proceeds from some unknown cause, or is an unusual effect of a known cause' (Oxford English Dictionary 1989). This implies that an accident is an event or occurrence, which may have a negative outcome of varying degrees of severity on the person(s) involved, directly affecting their health or experience of tourism, and in extreme cases, may cause trauma for those involved. Such events are often described as mishaps although in the tourism context such a definition may need some modification to incorporate the predictable nature of accidents resulting from the tourists' undue care and attention. The term 'mishap' is frequently used to avoid the term 'accident' which implies no foreseeability. But what explanations can be advanced to explain the causes of accidents?

According to Heinrich (1931 in Petersen 1978: 15), people cause accidents since

> the occurrence of an injury results from a completed sequence of factors,
> the last one of these being the injury itself. The accident which caused the
> injury is in turn invariably caused or permitted directly by the unsafe act
> of a person and/or a mechanical or physical hazard.

Petersen (1978) argues that accident causation operates like a domino effect, with five stages, which influence the outcome:

- Ancestry or the social environment may condition the context of the environment associated with the accident;
- The fault of a person;
- An unsafe act or condition;
- The accident; and
- Injury.

This approach is widely used in accident investigation to examine the factors and stages associated in the accident although not all stages will necessarily be involved in each case. In addition, by removing one of the above stages, it may be possible to break the domino effect by removing a sequence. However, in all cases there may also be the effect of multiple causes, such as contributory factors, causes and sub-causes that influence the context and features of individual accidents. For example, inadequate safety training for employees, weaknesses in management systems to prevent accidents, an absence of policies, procedures, training and enforcement of legislation may affect accidents (Reason 1990). Even so, Petersen (1978: 19) argues that it is no longer possible to predict the relationship between the severity and frequency of accidents because "originally, studies seemed to show that for every serious accident we can expect to suffer 29 minor accidents and 300 no-injury accidents or near misses".

It is now widely acknowledged that the relationship which exists will vary according to the particular human activity involved, and the circumstances causing a minor accident may be different to those affecting the severe accident. Therefore, in a tourism context accidents may occur for various reasons as the discussion of multi-causation theory implies. Cossar (1995) argues that travel exposes the individual to new, cultural, psychological, physical, physiological, emotional, environmental and microbiological experiences and challenges. The individual's ability to cope and survive such changes is affected by a variety of human factors such as their physical, mental and medical condition. Subsequently, this is also influenced by their behaviour, which is influenced by personality, experience, education, race, gender, age and social status. One of the key factors which social psychological studies of tourists have examined is the behaviour change when people are tourists, where "the relaxed attitudes and reduced inhibitions which are natural elements of vacational enjoyment expose the traveller to risks" (Cossar *et al.* 1990: 36) which one might normally avoid. This point is reinforced by Bewes (1993) where

> the danger of experiencing trauma during overseas travel . . . This is in
> part due to accidents during travel and in part to other dangers. In a great
> many countries overseas the possibility of experiencing a road traffic

accident is significantly greater ... road accidents are more dangerous ... although fatal air accidents attract much more publicity than road accidents (Bewes 1993: 454–455).

Although Bewes (1993) cites no international statistics, Dawood (1993: 281) argues that

Travellers need to be reminded that accidents represent the most significant hazard ... and that they are much more likely to die abroad from an accident than any other cause. Accidents abroad tend to follow a depressingly repetitive pattern. Injuries from moped accidents are especially common in young visitors to island resorts, for example. Renting a moped is cheaper than renting a car, and public transport is often poor; if it is hot, protective clothing and a helmet will be unattractive, even if they are available at all ... Motorists who wear seatbelts at home, use child seats for their children, observe speed limits and drink-drive laws, seem less inclined to do so abroad. And unsafe hotel balconies and swimming pools take their toll on travellers each year. Travellers should at least be alerted to the risks, and pressure be applied to tour operators who are in a position to make resort facilities more safe.

Therefore, many existing assessments of tourist accidents, even those conducted by leading researchers in the field, remain based on qualitative assessments and observations of common occurrences rather than quantitative evaluations of situations in individual countries.

While it is widely acknowledged in social psychology that health promotion cannot directly influence human behaviour and/or directly alter risk behaviour, it is recognised that individuals and groups need to have their awareness raised of specific issues (e.g. accidents). This is needed so that they can make an informed judgement to accept or reject the perceived risks involved with certain behaviour as tourists, a point highlighted above by Dawood (1993) and reiterated in Clift & Page (1996). Johnston (1989: 324) reviewed the literature on risk behaviour among tourists and recreationalists observing that

accidents are only one possible negative outcome of risk in mountains, and they accompany a sought-after possible positive outcome — satisfaction of motivations. Risk can be a motivation for recreation. Indeed, for some people the risk element that can be realised in certain activities and environments is the ultimate attraction. For others, risk is accepted as a necessary condition of such recreation in particular environments. And some others remain totally unaware of the risk element until they experience it.

Alternatively, for certain types of tourists seeking adventure tourism experiences, risk will be an inevitable trade off but for the majority of tourists, risk will not necessarily be so obvious because of the unknown nature of risks involving accommodation, driving on unfamiliar roads, swimming in the sea and participating in activities in unfamiliar environments. One outcome of Johnston's (1989) study is that the apportioning of blame for a high number of accidents is often based on misconceptions about the frequency of accidents, a feature reinforced by media reports.

A recent study by Girasek (2001) questions the premise "among the injury community . . . that injuries cannot be prevented." What Girasek (2001) confirmed in the case of the United States of America was that no national studies of unintentional injury existed, and so by implication, it is not that surprising if the accident literature in terms of tourist activity is scant. This is what makes an attempt to define the scale, extent and scope of tourist injury in one country, such as New Zealand, so significant in the research community. Girasek's (2001) analysis of injury-related prevention beliefs highlighted the questioning of the conventional attitudes that injury is in fact preventable and inevitable; that accidents just happen and there is nothing you can do to prevent them occurring. Increased understanding of risk behaviour and individuals' attitudes to accident preventability may make interventions from a public policy and health professional perspective more effective. Therefore, although disagreement may exist on whether accidents are random events or preventable, there is certainly a complete dearth of literature on tourism. This is because research is still establishing the extent and nature of the problem rather than moving on to the next stage of establishing what interventions and attitude change is necessary to make interventions effective.

Tourist Accidents in New Zealand

The Accident Rehabilitation and Compensation Act (1972) (New Zealand Statutes 1972) provides cover (commonly referred to as ACC cover) for overseas tourists on a no-fault basis during their visit to New Zealand. To receive cover (i.e. the cost of treatment for an accident), visitors must see a doctor and complete the appropriate claim form. Cover for overseas tourists include costs of medical treatment, hospital treatment, transport expenses and some rehabilitation costs.

Accident Rehabilitation and Compensation Insurance Scheme (ACC) Records as a Data Source

ACC records are a unique source of data. The scheme is one of only a few worldwide which specifically identifies overseas tourists as a client group and records only those claims where an entitlement was due. Where medical only claims are made (e.g. a tourist visits a general practitioner for treatment) these are not recorded in the data. The implications of this are that the total claims analysed in this study based on eligibility for cover may only deal with a small proportion of the total overseas tourist accidents. Moreover, overseas visitors may not register injury claims to ACC, as most come from countries in which medical and rehabilitation costs are not covered by a public insurance scheme (McKay 1998). ACC's Treasury Division provided statistical summary data for a range of variables related to tourist claims. Data from ACC claims were examined on the basis of the methodology and procedure followed by Page &

Meyer (1996). It was not possible to replicate all the analysis reported in this earlier study due to the ACC's trend to reduce the amount of information collated for each claim. This reduced data availability is concerning at a time of growing attention to tourist safety.

Analysis of ACC Data for Overseas Visitor Accidents

ACC's non-earner account records overseas tourist claims as these are recorded as non-work-related accident claims. Claims for the financial year 1995–1996, comparing non-earners and overseas visitors, are presented in Table 3.1. The incidence of claims relative to population size is also shown. The population of New Zealand was approximately 3.6 million for the year ending 30 June 1996. The average duration of stay for an overseas visitor was 19.3 days during this same period. By multiplying claim rates for overseas visitors by the factor 18.91 (365/19.3) comparable data was produced with those for the resident population of New Zealand.

The total claim rate of 5.0 claims per 1,000 overseas visitors per annum is well below the total claim rate of 14.0 for total non-work injuries. The claim rates for both groups are lower than those calculated by Page & Meyer (1996), which were 6.742 and 23.465 respectively, and follow a trend observed since July 1989. The comparison of claim rates for overseas visitors with non-earner claims suggests that overseas visitors are less likely to experience accidents than New Zealand residents. It should be noted, however, that many overseas visitors may be unaware of the existence of ACC, and may rely on their private insurance cover in the event of injury while in New Zealand. This in itself is a major weakness in the data. Furthermore, the mean claim rate is considerably higher for overseas visitors ($821) than for residents ($529), indicating that overseas visitors tend to register claims for more serious injuries. Conversely, it may indicate that residents tend to register claims for less serious injuries more often. Two activity categories, 'Recreation and Sport' and 'Travelling as a Passenger', reveal higher rates for overseas visitors (2.227 and 0.171 respectively) than for residents (2.217 and 0.012 respectively). Recreational and sporting activities also comprise over 44% of claims by overseas visitors, compared to approximately 16% of non-work injury claims.

The age distribution for claimants may be an important factor in explaining differences in claim rates between overseas visitors and residents. Table 3.2 compares age distributions for non-earners and overseas visitors for claims registered in the year ending 30 June 1996. The young and old register the majority of non-earner claims, with only 30.5% of residents' claims registered in the 20–64 age group. For overseas visitors, the largest claim rates are observed for the 20–34 age group, with 74.1% of claimants located in the 20–64 age group.

Examination of the geographical distribution of overseas tourist claims provided a further opportunity to consider the possible influence of adventure tourism on ACC claim rates for overseas tourists. A large proportion of accidents were located in the Otago region (29% of claims by overseas tourists), with this spatial concentration being consistent for the 1993–1996 period considered in the analysis. Adventure tourism activities are known to be spatially concentrated in the Otago region, particularly in and

Table 3.1: Non-work injury claims by activity of injured claimant for the year ending 30 June 1996 (ACC).

Activity	Total ACC claims Non-earners			ACC claims Overseas visitors		
	Number (%)	Rate (*)	Mean claim	Number (%)	Rate (**)	Mean claim
Ascending/descending	1352 (2.7)	0.377		4 (0.0)	0.052	
Children playing	5702 (11.3)	1.588		14 (3.6)	0.182	
Driving, riding (***)	8604 (17.1)	2.396		48 (12.4)	0.625	
Eating, drinking	541 (1.1)	0.151		3 (0.0)	0.039	
Fighting	1080 (2.1)	0.301		2 (0.0)	0.026	
Getting on/off, in/out of	1715 (3.4)	0.478		10 (2.6)	0.130	
Lifting, lowering	1476 (2.9)	0.411		1 (0.0)	0.013	
Recreation or sport	7962 (15.8)	2.217		171 (44.3)	2.227	
Travelling as a passenger	44 (0.1)	0.012		13 (3.4)	0.171	
Walking, running	11487 (22.8)	3.199		83 (21.5)	1.081	
Other/unclassified	10375 (20.6)	2.881		37 (9.5)	0.482	
Total	50338	14.017	$529	386	5.028	$821

(*) Rate per year per 1,000 population.
(**) Rate per year per 1,000 visitors with an average duration of stay equal to 19.3 days.
(***) Source for driving, riding: activity for non-earners' account entitlement claims plus non-work motor vehicle entitlement claims.

Table 3.2: Age comparison for claims registered by non-earners and overseas visitors for the year ending 30 June 1996.

Age at time of accident	Non-earner claims (%)	Overseas visitor claims (%)
0–14	16741 (38.6)	35 (9.1)
15–19	4988 (11.5)	43 (11.1)
20–24	2473 (5.7)	75 (19.4)
25–34	4074 (9.4)	109 (28.2)
35–44	2793 (6.4)	28 (7.3)
45–54	1674 (3.9)	42 (10.9)
55–64	2206 (5.1)	32 (8.3)
65 +	8457 (19.4)	22 (5.7)
Total	43406 (100)	386 (100)

around Queenstown (Berno & Moore 1996). This suggests that adventure tourism accidents are making a significant contribution to claims made to ACC. This is not surprising as Queenstown promotes itself as the adventure tourism capital of New Zealand with a wide range of adventure tourism activities available in the region. Further evidence for the role of adventure tourism in accident claims to ACC comes from an examination of the age of claimants who incurred accidents in the Otago region. Visitors in the 21–40 years age group represent the typical age profile of overseas visitors engaging in adventure tourism and registered 56% of claims. Finally, of overseas tourist claims located in Otago, 71% were related to recreation or sporting activities (adventure tourism activities are most likely to be recorded by ACC as recreation or sport).

Finally, the analysis examined trends in overseas visitor claims made to ACC and the cost of claims to ACC. Claim rate frequencies based on the total number of overseas visitors to New Zealand were calculated for the period 1990–1996, and are shown in Table 3.3. Claim rates can be seen to have fallen each year from 1990–1995, but to have risen slightly during 1995–1996. The extent to which these figures represent a real reduction in accident incidence is unclear, as they may to a large extent reflect changes in ACC legislation during 1992 regarding guidelines for cover and payments. The mean claim payment is much more irregular, with payments ranging from $1,169 to a high of $2,497. This irregularity has resulted in a similar pattern for total payments.

Complementary Data Sources on Tourist Accidents

No one body or organization in New Zealand is responsible for investigating tourism-related accidents or for collating statistics on accident incidence in the tourism industry.

Table 3.3: Frequency of claim rate for overseas visitors.

Year	1989–1990	1990–1991	1991–1992	1992–1993	1993–1994	1994–1995	1995–1996
Claims paid	497	497	483	453	493	417	493
Arrivals (000s)	920.6	914.0	986.0	1068.2	1196.8	1322.9	1437.1
Claims paid Per 1,000 arrivals	0.5398	0.5438	0.4899	0.4241	0.4119	0.3152	0.3431
Mean claim Payment (NZ$)	1891	2497	2104	1757	1169	1280	1966
Total payment (NZ$000,000)	0.940	1.241	1.016	0.796	0.576	0.534	0.969

Rather a range of agencies document accidents related to their area of legislative responsibility. These include the Department of Labour's Occupational Health and Safety Division, the Maritime Safety Authority, the Civil Aviation Authority, the Transport Accident Investigation Commission and the New Zealand Ministry of Health. A number of these sources are now outlined.

Mountain Recreation Related Fatalities

The New Zealand Mountain Safety Council undertook an analysis of fatalities among international and domestic visitors from Coroners' inquest reports for the period 1991–1993, thereby expanding the innovative research by Johnston (1989). From these, those cases of adventure tourism or individual adventure recreation were extracted. In 1991, there were 71 outdoor fatalities in New Zealand, of which eight fall within the broad category of adventure tourism, and just two involved a professional guide or company. In 1992, 7 of 80 fatalities were adventure tourism related, just one of which was guided. During 1993, 10 of 67 fatalities were adventure tourism related, four of which were guided. While data for more recent years were not available at the time of printing, personal communication with the Mountain Safety Council has indicated these trends have tended to persist. The majority of these fatalities occurred in the South Island of New Zealand, corroborating the geographical patterns observed for ACC claims. Mountaineering and tramping emerge as significant activities, reflecting the large number of people who visit locations such as Mount Cook and the Queenstown area. The age profile of tourists involved in fatalities also corresponds with the main client group for adventure tourism, with all but one fatality being aged between 19 and 36 years.

Maritime Accidents: The Tourist Dimension

The Maritime Safety Authority (MSA) investigates accidents reported to them under the Maritime Transport Act (1994). Water-based adventure tourism activities such as white water rafting, jet boating and water and sea kayaking fall within the MSA's jurisdiction. During the period September 1995–August 1996, 370 events were reported to the MSA, representing a substantial increase on the 280 accidents, incidents and mishaps reported the previous year. The increase represents the extent of under-reporting prior to the 1995 legislation, rather than a real increase in the number of events. In 1995–1996, seasonality in accidents led to peaks in January and February, coinciding with the summer season and a rise in recreational boating activity. In 1995–1996, there were 37 fatalities in the marine environment, with 17 deaths associated with pleasure boats, 11 while fishing, 6 on commercial boat trips such as rafting or jet boating. The MSA argues that only a small proportion of pleasure boat accidents are being reported.

   Table 3.4 summarises the results of a survey of MSA accident investigation reports where tourists were involved in some way. In the period between mid-1994 to the end of 1998, 43 investigations were undertaken with an adventure tourism component. A number of these were reported because of breaches of Maritime Transport legislation with no other incident or involved fairly minor non-hazardous events. These incidents are not included in Table 3.4. The 31 incidents included in the analysis do include a number of potentially serious near miss and non-injury incidents, however, based on access to the accident reports for each incident. It is apparent from Table 3.4 that white water rafting and jet boating were the most commonly reported adventure tourism-related accidents. These incidents ranged in severity from near- miss or minor injury incidents to fatalities, with white water rafting having by far the largest number of fatalities. Drowning was the cause of almost all fatalities, and a number of near-miss incidents involved serious non-fatal submersions. The number of incidents in which injuries were sustained ($n = 10$) does not include non-fatal submersions or extended exposure to the cold and wet, although these occurred fairly frequently in rafting incidents, and are recognised as negative and potentially fatal occurrences.

   Clearly, white water rafting incidents represent a major concern for the New Zealand tourist industry (Chou 1996). After publicity surrounding a series of white water rafting accidents in Queenstown in 1995 (McLaughlan 1995), a report commissioned by the MSA downplayed the perception of risk highlighted by media reports (CM Research 1995). The New Zealand Water Safety Council provides evidence of fatalities involving white water rafting over a longer period of time. In the period 1980 to 1995, 16 white water rafting participants lost their lives, five of whom were overseas visitors. These figures have led to the establishment of the New Zealand Rafting Association in January 1997 to supervise and control the industry. Further intervention to increase safety in the rafting and jet boating sectors of the adventure tourism industry has come in the form of Ministry of Transport's Maritime Rules (Part 80), for Marine Craft used for adventure

Table 3.4: Summary of maritime safety authority accident reports for adventure tourism incidents mid-1994 to end 1998.

Type of activity	Incidents	Fatalities	Incidents with injuries	Common incidents
White water rafting	8	4	5	• Capsizing • Slip or fall from raft
Jet boating/speed boats	13	1	3	• Mechanical failure • Collisions
Commercial dive boat expeditions	5	1	0	• Collisions • Capsizing
Dolphin watching/swimming trips	2	0	2	• Passengers falling onto deck
Kayaking/canoeing	2	1	0	• Capsizing
Caving/tubing	1	1	0	• Drowning
Totals	31	8	10	

tourism (rafts and jet boats). This legislation is made pursuant to Section 36 of the Maritime Transport Act 1994.

Tourist Scenic Flight Accidents

Scenic flights offer visitors the opportunity to land on scenic attractions such as glaciers at popular tourist locations such as Mount Cook and other destinations throughout New Zealand (see Page 1999 for further detail). The New Zealand Civil Aviation Authority (CAA) produces statistics for aviation-related accidents, although the data are such that they do not allow analysis to determine the involvement of overseas tourists. According to CAA reports, there were 235 aviation accidents in the years 1992–1995, less than 10% of which were fatal. National figures for air crashes in New Zealand were published in the New Zealand press (Sunday Star-Times April 25 1999). These figures claim a total of 1,604 air crashes have been reported since 1985, approximately 10% of which involved fatalities ($n = 157$), with a total loss of life of 311. Unfortunately, it is not clear what proportion of these incidents involved tourist scenic flights. Page & Meyer (1997) provided a more useful account of the extent of overseas tourist involvement in fatal scenic flight incidents. They reported that 27 overseas tourists were killed in five scenic flight accidents between October 1986 and October 1993. A further six domestic tourists died in two scenic flight incidents during this period. The most common mode of transport involved in scenic flight fatalities was the helicopter.

Hospitalised Admission of Tourists

The Ministry of Health's New Zealand Health Information Service (NZHIS) public hospital morbidity data files routinely collated on all discharges from public hospitals provide Accident and Emergency facilitates for the population. The data set contained information about all non-resident discharges from public hospitals throughout New Zealand for the period 1982 to 1996 (see Bentley *et al.* 2001). The large majority of cases included in the analysis were tourist injuries, with a small but unknown proportion of cases involving non-residents working off-shore (mostly fishing and other vessel crew) and persons working in New Zealand on overseas visitor work permits.

The circumstances of injuries are coded according to the International Classification of Diseases (World Health Organization 1978). The numerous three digit E-code values found within the data set were collapsed into 17 event groups, providing a variable more suitable for analysis. Adventure tourism activities were identified from content analysis of the 'one-line' event descriptions. Therefore, for relevant event group categories (i.e. motor vehicle non-traffic accidents; cycles; animals; watercraft; aviation; falls from a height and on the level; struck by/strike against object or person), the 'one-line' descriptions of accident circumstances were examined to determine the involvement of recreational/adventure activities.

The spatial distribution of overseas visitor injuries within New Zealand had to be determined indirectly from information on the hospital to which the injured person was admitted assuming admission to the nearest regional hospital. Data on the nationality of injured overseas visitors was insufficiently complete to undertake a worthwhile analysis, which was a significant limitation of the study. Some 5,863 overseas visitors were admitted to New Zealand public hospitals due to injuries sustained during the 14 year period 1982–1996. Injury incidence rates (per 100,000 overseas visitor arrivals) were calculated for each year 1982–1996 and are shown in Table 3.5.

The overall incidence rate of 44 injuries per 100,000 overseas visitor arrivals indicates that approximately one in every 2,300 overseas visitors was admitted to hospital with an injury during their stay in New Zealand. One half of visitors were hospitalised for a period of two days or less, while 80% spent less than 10 days in hospital and only 10% of injuries required more than 15 days hospitalisation. These findings were reflected in the distribution of injury severity scale classifications with 17% of cases classified as 'serious', and 1% as 'severe'. The majority of cases were classified as 'moderate' injuries (51%).

Fifteen percent of overseas visitor injuries were sustained at a 'place for recreation and sport', with other significant injury locations being 'street or highway' (13.5%), 'home' (7.1%) and 'public building' (4.4%). In a large number of cases ($n = 2,785$,

Table 3.5: Overseas visitor injury incidence 1982–1996.

Year	Overseas visitor arrivals ($\times$ **1000**)	Hospital admissions		Injury Incidence Rate*
		n	%	
1982	472.6	173	3.0	37
1983	487.7	245	4.2	50
1984	518.4	243	4.1	47
1985	570.0	291	5.0	51
1986	689.1	304	5.2	44
1987	763.2	379	6.5	50
1988	855.5	310	5.3	36
1989	867.5	383	6.5	44
1990	993.4	444	7.6	45
1991	967.1	423	7.2	44
1992	999.7	492	8.4	49
1993	1086.6	589	10.0	54
1994	1213.3	606	10.3	50
1995	1343.0	454	7.7	34
1996	1441.8	527	9.0	37
Total	13268.9	5863	100	44

* (per 100,000 arrivals).

Table 3.6: The role of adventure tourism in overseas visitor injury morbidity 1982–1996.

Event group	*n*	(%)	Adventure tourism activities*	*n*	(%)	(%) all cases
Motor Vehicle Traffic	1604	27.4				
Motor Vehicle Non-Traffic	59	1.0	Quad/farm bikes	*18*	*30.5*	*0.3*
Pedal Cycle	165	2.8	Road cycling	*95*	*57.6*	
			Mountain biking	*13*	*7.9*	*0.2*
			Other/unclassified	*57*	*34.5*	
Animal-related	174	3.0	Horse (fell from)	*153*	*87.9*	*2.6*
			Horse (kicked by)	*18*	*10.3*	*0.3*
			Bull (rodeo)	*3*	*1.8*	*0.05*
Watercraft-related	320	5.5	White water raft	*46*	*14.5*	*0.8*
			Jet boat	*21*	*6.6*	*0.4*
			Kayak/canoe	*3*	*1.0*	*0.05*
			Diving	*3*	*1.0*	*0.05*
			Crew/fishing boat	115	36.0	
			Unspecified boat/ship	132	41.6	
Aviation-related	100	1.7	Parapenting/gliding	*27*	*27.0*	*0.5*
			Skydiving	*23*	*23.0*	*0.4*
			Glider/unpowered	*5*	*5.0*	*0.08*
			Hang glider	*4*	*4.0*	*0.06*
			Crew/work-related	4	4.0	
			Unspecified aircraft	24	24.0	
			Helicopter	13	13.0	
Falls from a height/ Falls on the same level	2027	34.6	Skiing/S.boarding	*344*	*17.0*	*5.9*
			Mountaineering/ Tramping	*260*	*12.9*	*4.4*
			Luge	*24*	*1.1*	*0.4*
			Flying fox	*18*	*1.0*	*0.3*
			Parapenting	*9*	*0.4*	*0.2*
			White water rafting	*5*	*0.2*	*0.08*
			Playground activity	40	2.0	
			Swimming pool/spa	27	1.3	
Struck by/strike against object or person	325	5.5	Skiing/Snowboarding	*20*	*6.2*	*0.3*
			Mountaineering/ Tramping	*10*	*3.1*	*0.2*
			Rugby/other sports	108	33.2	
Other (non-recreational events)	1089	18.6				
Total	5863	100	*Total estimated adventure tourism*	*1027*		*17.5*

* Adventure tourism activities determined from content analysis of 'one-line' descriptions of accident circumstances provided in narrative fields. Adventure tourism activities shown in italics.

47.5%) the injury location was coded as 'unspecified place'. Based on the assumption that 15% of unspecified cases were recreational, it is estimated that the actual proportion of overseas visitor injuries occurring in a 'place for recreation and sport' is likely to be approximately 20%. This suggests an overall sport and recreation injury incidence rate of approximately nine injuries per 100,000 overseas visitors.

Analysis by *Event* (determined from analysis of three digit event description E-codes and information extracted from narrative fields) allowed for a more detailed identification of the role of recreational activities in overseas visitor injury (Bentley *et al.* 2001). Table 3.7 outlines each 'event group' and the recreational activities within each event group. The largest event group category was falls (from a height and on the level), followed by motor vehicle traffic accidents (see Page *et al.* 2001 for a discussion of this theme). The distribution of adventure tourism activities within 'falls' and other non-traffic event groups indicate that these form a significant contribution to hospital

Table 3.7: Place of occurrence categories by geographic region.

Region	Total		Sport and Recreation	Road	Home	Public building	Other
	n	%	%	%	%	%	%
North Island							
Northland	400	6.8	9.4	28.1	11.2	3.9	47.5
Auckland	1358	23.2	9.3	13.7	15.0	6.7	55.3
Waikato	233	4.0	16.7	33.5	13.5	1.6	34.7
Bay of Plenty/Coromandal	173	3.0	14.8	16.3	20.7	1.5	46.7
Eastland/							
Rotorua	416	7.1	36.7	16.1	6.8	12.0	29.4
Central NorthIsland	138	2.4	26.3	13.2	11.0	5.5	44.0
Hawkes Bay	119	2.0	14.3	28.6	8.6	4.3	44.2
Wairapapa/	197	3.4	18.8	17.3	10.5	6.8	46.6
Wanganui/New Plymouth							
Wellington	112	1.9	12.1	12.1	15.7	6.0	54.1
South Island							
Marlborough/Abel Tasman	210	3.6	8.1	21.0	12.8	4.7	53.4
West Coast	267	4.6	8.7	31.6	1.0	5.1	53.6
Canterbury	760	13.0	25.5	14.4	5.7	5.5	48.9
Otago	934	15.9	35.5	12.1	6.7	6.6	41.4
Southland/	483	8.2	26.2	27.7	2.1	4.4	39.6
Fiordland							
Other	63	1.1	12.0	30.0	10.0	0.0	48.0
Total	5863	100					

  Notable sport and recreation contribution to regional injuries.

injury admissions among overseas visitors. The major recreational activities were skiing/snowboarding (6.2% of all injury cases), mountaineering/tramping (4.6%), horse riding (2.9%), and cycling (2.8%). Water-based activities included white water rafting (0.8%) and jet boating (0.4%), although the proportion of adventure tourism-related injuries involving 'unspecified boats/ships' cannot be determined. Aviation-related activities included parapenting/gliding (0.5%) and skydiving (0.4%). The incidence of scenic flight injuries among helicopter and aircraft cases is not documented. It is pertinent to recognise that a significant number of injuries affect recreationalists and adventure tourists who are not taking part in organised commercial adventure tourism activities (i.e. mountaineering/tramping/skiing).

The total contribution of adventure tourism related activity to overseas visitor injuries for the years 1982–1996 comprised 1,027 events. These consist of 17.5% of all overseas visitor injuries. This total probably under-represents the true involvement of adventure tourism activity in overseas visitor injury due to the unknown extent of unclassified/unspecified events which may have been adventure tourism-related. Based on the assumption that 17% of all such events were adventure tourism-related, an adjusted total of 1,109 adventure tourism-related events were obtained, representing approximately 19% of overseas visitor injuries. This indicates an overall adventure tourism-related injury-incidence rate of approximately eight injuries per 100,000 overseas visitors. This contribution to overseas visitor morbidity is considerable, as demonstrated when weighed against that for motor vehicle traffic accidents (12 injuries per 100,000 overseas visitors), an activity for which there is significantly higher exposure amongst the tourist population.

The spatial distribution of overseas visitor injuries within New Zealand was determined indirectly from hospital codes (denoting the location of the hospital the injured person was admitted to). Table 3.7 outlines geographic region and major place of occurrence.

The region with the largest number of overseas visitor hospital admissions was Auckland (23%), followed by Otago (15.9%), Canterbury (13%) and Rotorua (7%). This reflects the typical pattern of movement throughout New Zealand by international visitors contained in the latest International Visitor Survey (Page & Thorn 1997; Page & Hall 1999). The high proportion of Auckland hospital admissions is probably a reflection of the fact that Auckland is the main gateway to New Zealand (Page 1999; Page & Hall 1999). From the place of occurrence categories data, the higher proportions of sport and recreation injuries occurred in those regions known to be the major centres for adventure tourism activity in New Zealand: Rotorua, the Central North Island, Canterbury, Otago and Southland. In support of the role of adventure tourism in overseas visitor injury incidence, 8% of all cases were admitted to hospitals in the Queenstown (Central Otago) area.

Almost one-half of injuries to overseas visitors occurred during the main summer months of December (11.7%), January (12.2%), February (12.3%) and March (11.2%). Monthly injury-incidence rates (per 100,000 overseas visitors) were determined for the years 1989–1996. Highest injury rates were observed for the peak summer months, with injury rates of at least five per 100,000 overseas visitors for January, February and March. Interestingly, injury-incidence rates were significantly lower for the month of

December (3.6). A second peak in injury-incidence was observed for the winter months, July (4), August (4.5) and September (4). Lowest overseas injury-incidence rates were observed for May, October and November (all 3.1). Twenty-two percent of accidental deaths to overseas visitors involved some form of adventure tourism ($n = 99$). Activities most commonly involved in fatal accidents included mountaineering and tramping ($n = 44$), scenic flights ($n = 25$) and cycling ($n = 10$).

Conclusion

This chapter highlighted the issue of accidents among overseas tourists and deficiencies in the data that documents tourist injuries and accidents. In many cases data sources not associated with tourism offer a number of insights from secondary data sources. Combining the various secondary sources to present a baseline of data is the initial stage in examining tourist accidents and injuries in a sustained and proactive manner. To derive a complete pattern of injuries for tourists requires researchers to collaborate and cooperate in data sharing, critical debate and the culmination of multidisciplinary research. This reflects the problem in most countries that no one agency or body is responsible for monitoring or investigating tourism accidents. A more holistic assessment of tourist safety and well-being may be the next step to assess how the accident process is associated with tourist behaviour and activity patterns.

Primary research, which examines tourist perceptions of risk, danger and safety, may well be the next stage of the research process to understand the psychology of tourist risk. Questions which need to be addressed by further research, include which tourism activities pose a greater accident and injury risk to overseas visitors than to residents, and the extent to which overseas visitors are more or less at risk participating in adventure tourism than during other more passive activities. To fully investigate these issues requires a more detailed behavioural analysis of tourists, their activity patterns, the impact of marketing and promotion on risk-taking, and the extent to which risk is adequately/inadequately explained to participants. There is also a need for greater attention to the development of improved risk management systems for adventure tourism operations.

References

Bentley, T., Meyer, D., Page, S. J., & Chalmers, D. (2001). Recreational tourism injuries among visitors to New Zealand: An exploratory analysis using hospital discharge data. *Tourism Management, 22*, 373–381.

Bentley, T., & Page, S. J. (2001). Scoping the extent of tourist accidents in New Zealand. *Annals of Tourism Research, 28* (3), 705–726.

Berno, T., & Moore, K. (1996). *The nature of the adventure tourism experience in Queenstown.* Paper Presented at the Tourism Down Under Conference, Centre for Tourism, University of Otago, December 1996.

Bewes, P. (1993). Trauma and accidents. In: R. Behrens, & K. McAdam (Eds), *Travel medicine* (pp. 454–466). London: Churchill Livingstone.

Chou, R. (1996). *Managing risk in New Zealand adventure tourism: A case study of commercial white water rafting.* Unpublished Research Report for the Master of Business Studies, Massey University at Albany, Auckland.

Clift, S., & Page, S. J. (Eds) (1996). *Health and the international tourist.* London: Routledge.

CM Research (1995). *White water rafting customer research: Qualitative and quantitative research findings.* Report prepared for the Maritime Safety Authority White Water Rafting Safety Advisory Group. Wellington, New Zealand: CM Research.

Cossar, J. (1995). Travellers health: A medical perspective. In: S. Clift, & S. J. Page (Eds), *Health and the international tourist* (pp. 23–43). London: Routledge.

Cossar, J., Reid, R., Fallon, R. J., Bell, E., Riding, M., Folle, E., Dow, B. C., Mitchell, S., & Grist, N. (1990). A cumulative review of studies on travellers, their experiences of illness and the implications of these findings. *Journal of Infection, 21,* 27–42.

Dawood, R. (1993). Preparation for travel. In: R. Behrens, & K. McAdam, (Eds), *Travel medicine* (pp. 269–284). London: Churchill Livingstone.

Girasek, D. (2001). Public beliefs about the preventability of unintentional injury deaths. *Accident Analysis and Prevention, 33* (4), 455–465.

Hargarten, S. W., Baker, M. T., & Guptill, K. (1991). Overseas fatalities of United States citizen travelers: An analysis of deaths related to international travel. *Annals of Emergency Medicine, 20,* 622–626.

Johnston, M. (1989). Accidents in mountain recreation: The experiences of international and domestic visitors in New Zealand. *GeoJournal, 19* (3), 323–328.

McKay, D. (1998). *Don't run the risk.* New Zealand Local Government, May, pp. 39–41.

McLaughlan, M. (1995). White water death: Why is the shotover New Zealand's most lethal river? *North and South,* (December), 70–81.

New Zealand Statutes (1972). *Accident Compensation Corporation Act 1972.* Wellington: Government Print.

Nicol, J., Wilks, J., & Wood, M. (1996). Tourists as inpatients in Queensland regional hospitals. *Australian Health Review, 19* (4), 55–72.

Oxford English Dictionary (1989). Oxford: Oxford University Press.

Page, S. J. (1999). *Transport and tourism.* Harlow: Addison Wesley Longman.

Page, S. J., & Hall, C. M. (1999). New Zealand. *Travel and Tourism Intelligence Country Report, 4,* 47–76.

Page, S. J., & Meyer, D. (1996). Tourist accidents: An exploratory analysis. *Annals of Tourism Research, 23* (3), 666–690.

Page, S. J., & Meyer, D. (1997). Injuries and accidents among international tourists in Australasia: Scale, causes and solutions (pp. 61–79). In: S. Clift, & P. Grabowski (Eds), *Tourism and health: Risks, research and responses.* London: Pinter.

Page, S. J., & Thorn, K. (1997). Towards sustainable tourism planning in New Zealand: Public sector planning responses. *Journal of Sustainable Tourism, 5* (1), 59–77.

Page, S. J., Brunt, P., Busby, G., & Connell, J. (2001). *Tourism: A modern synthesis.* London: Thomson Learning.

Paixao, M. L., Dewar, R., Cossar, J., & Reid, D. (1991). What do Scots die of when abroad? *Scottish Medical Journal, 36* (4), 114–116.

Petersen, D. (1978). *Techniques of safety management.* Tokyo: McGraw Hill.

Reason, J. (1990). *Human Error.* New York: Cambridge University Press.

Wilks, J., Pendergast, D., & Service, M. (1996). Newspaper reporting of tourist health and safety issues. *Australian Leisure, 7* (3), 45–48.

World Health Organization (1978). *International classification of diseases* (9th rev.). Geneva: World Health Organization.

Section 2

Adventure tourism

Chapter 4

Risk Acceptance in Adventure Tourism — Paradox and Context

Chris Ryan

Ewert (1989: 8) defines adventure tourism as the "deliberate seeking of risk and the uncertainty of outcome". This chapter argues that the definitions of risk and uncertainty are themselves complex concepts mediated through socio-psychological constructs. While other chapters in this book are concerned with risk management within site-specific locations or other specificities of risk measurement, risky behaviours or risk prevention, this chapter seeks to contextualise risk acceptance by tourists partaking in adventure tourism by reference to broader socio-psychological factors. It argues that context, perceived risk and benefits need to be incorporated into any model to understand why tourists are prepared to take risks on holiday that they may not normally accept in everyday life. The chapter begins by considering broad social conditions, presents a hypothesis that has been suggested by some commentators, and then seeks to modify that contention by reference to published research. The role of the adventure provider in defining acceptable risk is also addressed.

Within wider society, risk analysis, its measurement and role within scenarios like disaster planning and mitigation have become increasingly common. Among the phenomena that have been studied are risks associated with earthquakes, nuclear power stations, flooding and environmental pollution resulting from accidents at sea and other locations. It might be observed that as technology becomes increasingly sophisticated and thus protects humans from risks that once threatened communities, so new risks in turn become associated with the new technologies. At one level, the actuarial, it might be said that the measurement and understanding of risk has become better understood.

Underlying this approach is the view that risk is something to be avoided, or at least mitigated, and thus the term is associated with the words 'dangerous' and 'hazardous'. Certainly it appears that the etymological roots of the word contain these nuances. Bernstein (1996) however, argues that the word originates from the Italian '*risicare*' that appeared in the seventeenth century, and that it can be translated as meaning 'to dare'. Alternatively, Keey (1998) holds that it means 'to run into danger'. On the whole there appears to be a consensus that the word first developed a common usage as the need for,

Managing Tourist Health and Safety in the New Millennium
Copyright © 2003 by Elsevier Science Ltd.
All rights of reproduction in any form reserved.
ISBN: 0-08-044000-2

and provision of, insurance became more common with the growth of sea-borne trade in the sixteenth and seventeenth centuries. Given this usage, the understanding of something being at risk and thus creating the probability of financial loss becomes understandable. However, within several forms of tourism, including adventure tourism, the concept of risk in its wider, less narrow interpretation as being something that is dared in order to gain advantage, something wherein the probability of loss is set aside in order to derive benefit, is thought appropriate. Elms (1998) suggests that risk possesses three related contexts, and these are shown in Figure 4.1. They are context, likelihood and consequence.

While in commercial business the consequence of risk is often stated to be potential loss, this is, of course, not wholly true. Indeed in classical economics, the very fourth factor of production after land, labour and capital is stated to be the entrepreneur who is rewarded with profit. So too, with the holidaymaker who engages in risk, except, as will be further discussed below, the reward is often of a psychological nature. The act of risk taking is undertaken with the view of obtaining a state of exhilaration, an adrenalin rush, or a state of well-being that will often have both immediate and possible long-term benefits.

The context of risk taking during holidays is also of interest, because, by definition, that holidaymaker is 'out of context'. Indeed, the holidaymaker might be said to be a dis-placed person from many perspectives — spatially in terms of not being within their home milieu (but equally of not being part of the host society) — psychologically given a potential predisposition toward the hedonistic or loss of normal responsibility — temporally as the role of holidaymaker is conventionally marginal to the life of the participant as a worker — socially as the holidaymaker enjoys a socially condoned period of non-work as a reward for work in a society where status is often associated with occupation. The context of risk taking on holiday is thus a specific social construct as well as being simply one of activity and place.

Given this context, the question of 'likelihood' begins to emerge as a complex one. There is first the technical component of possibility, of assessing the probabilities of an action or event occurring. But while this 'objective' (or actuarial) probability exists, there is also the perceived risk to be considered. It should be noted that technically 'risk' is not simply the probability of an event happening, but also the magnitude of loss (or

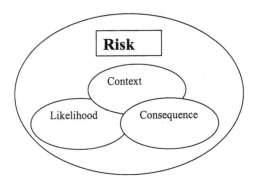

Figure 4.1: Components of risk.

gain) that is involved. Given that likelihood contains a perceptual component, particularly in holiday circumstances where the tourist as participant in an activity often will not have access to technical information about risk or hazard, then the social construct of risk within the holiday context becomes important. In short, 'likelihood' is as much a social-psychological construct as it is a statistical assessment — indeed arguably it is more the former than the latter under these circumstances of holidaying. As such, it is an outcome of judgement on the part of the holidaymaker — a judgement that incorporates evaluations of the guides, tour operators and even fellow holidaymakers as to their competencies and abilities. In as much as it is a judgement of a 'service,' then the services literature of the ServQual model implies that a judgement of tangible equipment, reliability, empathy, assurance and responsiveness to client needs are being made by the holidaymaker.

Accepting this approach implies that risk taking is the exercise of choice and thus the tourist-consumer is choosing to accept a risk in order to gain a benefit. It also implies that the tourist is assuming that the benefits of risk-taking are far more likely than the risk of loss. Certainly in the case of many adventure tourism products apparent adverse consequences of risk taking are serious. Indeed, loss of life is possible, as is demonstrated by the history of white-water rafting and associated products such as sledging down white waters in gorges. However, in spite of well publicised accidents, the statistical odds of death are very, very small, although literature exists about injuries suffered by tourists undertaking sporting activities like skiing. Indeed Bentley *et al.* (2000) comment that while the risk of serious injury is very small, higher incidences of minor injuries such as sprains and bruising tend to be associated with activities that involve the risk of falling from a moving vehicle or animal. They cite activities such as cycle tours, quad biking, horse riding and white water rafting as being those most likely to result in injuries to participants. However, it is also equally evident that tourists may suffer above average levels of risk in more mundane matters such as driving or even staying in hotel rooms where incidents of death through asphyxiation due to poorly maintained gas heaters have been recorded.

Given the nature of risk taking in many adventure tourism products and the apparent nature of risk that is involved a number of questions arise. These include:

- Why does such risk taking appear attractive?
- What is it about holidays that enhance risk taking?
- What level of knowledge exists among holiday-takers about risk associated with various activities?
- How are risks assessed?
- Are there any specific personality characteristics discernable among 'risk takers'?

So, why is it that risk-taking of this nature appears attractive? Why should tourists who often may simply engage in risks of normal urban living seek to place themselves in positions of discomfort, dependency on others and seek to use low levels of physical skills in natural settings? These three components of comfort, dependency and physical skills, it is argued, are not unrelated to the benefits that are being gained, and the socio-psychological contextualisation provided by western urban, technological societies.

Bernstein (1996) and Giddens (1998) have separately commented that modern notions of risk can only arise where society seeks to exercise control over the future. Paradoxically, as Giddens (1998) notes, the very use of technology that protects humans from natural disasters that arise from famine, flood and the like, has created in turn potentially more harmful human induced dangers that range from global warming to nuclear winters. But equally, or so this it is contended for the purpose of this chapter, the nature of individual risk-taking has also changed. Thus modernity both protects individuals from risks of nature and creates new types of technological and social risks. But modernity goes beyond this at the individual level. By reason of the physical protection offered by technology, modernity creates for many individuals, a psychological tension between the clear uncertainties of simple natural based risk that challenged people as individuals, and the unclear complexities of socially based risk that might not directly threaten individuals physically even while they affect patterns of social cohesion, partly through threats to psychological well-being, status and self esteem. Nor is this a simple matter of male testosterone, in that female clients of many adventure tourism products are as numerous as their male counterparts. In short, this feeling is experienced by both males and females, especially those engaged in more cerebral occupations. These concepts can be explored further.

For many people today, the physiological challenges associated with labour have been replaced by issues of social complexities wherein decisions made by individuals are in terms of assessing issues to tender advice or to contribute to wider discussions within organisations. Subsequent group-based decisions then made may take some time to deliver outcomes. The plaudits or blame for outcomes may be associated with organisations rather than individuals; and even then, at the individual level, outcomes may be measured more in terms of potential gains in career moves and/or status rather than any simple physiological outcome. Where physiological outcomes are experienced they are indirect, being mediated by feelings of depression, frustration or the like. So the nature of risk facing individuals in daily life has changed over the centuries to the point where, while today daily work life may indeed be stressful, (and indeed commuting to work by car is not without some dangers), the patterns of physical risk taking are less prevalent, less apparent and less immediate for many urban based occupations than might have been the case in the past. It is possible to contend that given this situation many will then seek compensating risk taking; which hypothesis implies that a psychological need for physical risk taking exists. Awareness of such needs implies action through deliberate choice. For their part, Lash et al. (1996: 13) have argued, "more and more areas of life . . . have been taken from the sphere of the natural and inevitable and made the objects of choice and responsibility".

Carr (2001) provides evidence for the thesis of the linkage between occupation and adventure in a survey of guided mountain climbers. Thus she cites one of her respondents as commenting that "being in the mountains gives me a challenge that my profession lacks. I can test my boundaries — the higher (the) difficulty the higher degree of personal satisfaction" (Carr 2001: 165–166).

Certainly it can be argued that the appeal of adventure tourism might lie in the immediacy of the situation where the complexities of modern life are replaced by situations that require instant, almost reflexive actions, and where the consequences of action are also immediate and appealing. In Skinnerian terms, action and effect, whether of reward or punishment, are both obvious and immediate; thereby creating clear and simple learning situations. Is there any evidence that such a proposal that risk taking within adventure tourism, and hence the appeal of adventure tourism, is due to changing social contexts? The evidence might be both direct and indirect. Perhaps it also needs to be noted that within the field of adventure tourism, almost by definition, there exists an inherent bias toward more educated and potentially, if not actually, higher income groups, as is generally found in studies of the socio-demographics of eco- and adventure tourists. (Students may have currently low incomes, but often possess above average earning potential). For example, Henderson & Ainsworth (2001) clearly show that opportunities and perceptions of leisure are socially and culturally bound. They found lower rates of leisure participation in physical leisure among African American and American Indian women in a cross-cultural activity participation study with the United States of America. One consequence of this is lower levels of good health are found among such groups. Similar findings are found (with much larger samples of 3,861) among migrants to Sweden and a linkage between physical activities and cardiovascular disease was found between different genders and ethnic groupings by Lindstroem & Sundquist (2001). Equally, studies of the economically deprived show that participation in physical activity helps sustain mental well-being (Pondé & Santana 2000). It can thus be concluded that predisposition to risk taking within adventure tourism type activities can be associated with certain income and ethnic groupings, with, as is evidenced by socio-demographic profiles of tourists, a tendency toward younger, western orientated, better educated groups tending to value the product, and thus accept the risks, more than other groups of people.

There is indirect evidence that boredom is a factor that predisposes people to risk taking; which risk taking is not always socially acceptable. For example Chang *et al.* (2001) in a study of young Australians found that those under the age of 18 would often cite boredom as a major reason for engaging in actions that could lead to petty criminal actions, or to more serious behaviours including drug taking. There is certainly medical evidence to suggest that passivity is associated with indices of poor health such as obesity, cardiovascular problems and high levels of cholesterol at statistically significant levels (e.g. see the work of Fung *et al.* 2000). It might be argued that as organisms seek to maintain health, then there is a predisposition to engage in physical activity that, by its very nature, implies the possibility of some risk, even at the comparatively minor level of, as noted by Bentley *et al.* (2000), falling over and suffering bruising, cuts and sprains. Within this scenario there is some evidence that personality has a role to play. Gibson (1996) found some evidence for the existence of a thrill-seeking personality in that of a sample of 1,277, some 124 males and 107 females were found to take thrill seeking vacations; that is, specifically seek out the products of the adventure tourism industry. However, what is of particular interest in Gibson's work is that the role of thrill-seeker seems associated with life stage, in that it is observed that the role peaks in early adulthood and thereafter declines. If this is the case then it implies that the socialisation processes referred to above wherein which risk taking is associated with frustrations of working life would need to be modified as being felt more keenly by younger adults or by certain personalities.

That this may well be the case is exemplified by a large number of studies. To cite but a few, McGuiggan (1999) applied the Myers-Briggs type indicator and found distinct responses based on factors like risk, preference for competition and pace of activity, although in this case it must be noted that the sample was small ($n = 103$). The research undertaken by Howard *et al.* (2002) is of interest through its exploring themes inherent in the work of Gibson (1996) and McGuiggan (1999). Creating scenarios within which to study four meta-motivational states, Howard *et al.* (2002) concluded that negativistic (as opposed to conventional) frames of mind were more important than arousal seeking when considering risk-taking behaviours. While these researchers were seeking to generalise results to broader social actions such as taking risks while driving or in sexual behaviour, of interest to concerns about adventure tourism is that one of the two specific experimental setting in this study was that of rock climbing. Farthofer and Brandstaetter (2001) argued, based on a small study of 20 steel workers, that subjective well-being among extroverts were personality characteristics that predisposed people toward high risk taking within leisure activities. It is suggested that this line of thinking is important. If the Howard *et al.* study (2002) is to be applied to adventure tourism, then it is not negativism, as it would conventionally be understood that is a determinant of risk taking, but a stance of non-conformity. Indeed, the work of Wang *et al.* (2001) specifically argues from a comparison of 23 patients with chronic primary insomnia and 28 'healthy' patients that, as measured by the Zuckerman Sensation Seeking Scale and the Zuckerman-Kuhlman Personality Questionnaire, thrill and adventure seeking is negatively associated with depression and feelings of negativity in the sense of neuroticism-anxiety measures. Indeed it is suggested that positive senses of self-worth have physical attributes in terms of healthy neurotransmitter systems and thalamic neural 'circuits'. However, a slightly contrary view is expressed by Carnelley & Ruscher (2000) who, based on a sample of 148 college students, found that those engaging in thrill and adventure seeking were, at least in part, motivated by a wish to obtain social approval and in fact tended to score highly in anxiety avoidance scales.

Turning to data derived from those actually participating in adventure tourism products then, from evidence derived from an analysis of the language used by participants at an adventure tourism location in New Zealand, Ryan & Ruthe (2002) would appear to endorse the view that the social atmosphere that is found in these locations, and the sharing of a common experience, would be a significant contributor to the experience of, in this instance, white water rafting. Arnold *et al.* (1998) studied white water rafting experiences on the Colorado-Utah border, and in their study they utilised semiotic labels of the cultural invention of wilderness, wilderness as a transcendental force, as having a restorative healing power, and wilderness as the last refuge that was essential to be conserved and preserved. Thus, from the perspective of this study, risk was secondary to the benefits being gained from participation. Walle (1997) specifically examines the relationship between the seeking of risk on the one hand, and self-fulfilment on the other. He argues that one view often postulated is that the outcome of participation in adventure tourism is a 'peak experience' akin to the process of self-actualisation described by Maslow (1954) in the latter's well-known hierarchy of needs or motives. Walle (1997) further proposes that to understand adventure tourism a second component is required, namely nature as a transcendental force. Participation therefore, it is suggested, is about developing experience and insights that are tested through activity. For Walle (1997: 277) "As a result of the focus towards risk ... the lucrative market segment which seeks insight seems to be under-served ... This under-served segment seeks adventure in order to interact with nature to gain insight, not to experience risk".

It is questionable to what degree participants in adventure tourism actually seek risk. Ryan (1997) adopts Csiksentmihayli's (1975) concept of flow to analyse the role of guides in a white water rafting venture. Based in part on the author's own experiences as a windsurfer of many years, including instruction in this sport, it is here suggested that what participants primarily require is a 'flow' experience wherein levels of skill and levels of challenge are congruent, thereby creating an unique, singular experience that is all consuming at the time. From this perspective risk may be defined as being that which is acceptable in as much as the participant feels able to cope with and manage the environment without detriment to one's own safety, and unacceptable limits of risk where the conditions are such as to threaten to overwhelm the participant are avoided. It follows from this that the assessment of risk is in part subjective but also in part measurable based on experience. It also means that risk is not inherently objective in the environment itself, but needs to be related to the competencies of the participant. Ryan's (1997) model further argues that not even those competencies are fixed because the experienced instructor or guide is able to direct, guide or instruct the participant to perform at levels higher than might otherwise be the case if the participant were left to their own devices.

From this literature review a number of conclusions appear to emerge. Figure 4.2 attempts to summarise the relationships being envisaged. First, any perception of risk is not simply 'objective' or 'actuarial'. The perceptions of risk are associated with a participant's level of skill, their level of knowledge about an activity, concepts of what

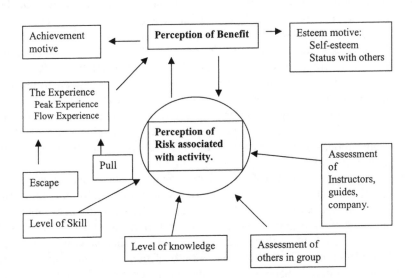

Figure 4.2: Risk and benefits in Adventure Tourism.

constitutes a risky situation, assessment of others in the group and trust in the level of guidance being provided by guides, instructors and others perceived to hold such positions. This shifts the understanding of risk from an 'absolute' into a 'relative' notion. For example, 'high' levels of risk may be anticipated by both the Hong Kong Chinese female office worker in terms of going on a soft adventure product because of the unfamiliarity of the terrain and the activity (e.g. canoeing down a river) and the young brash male who is engaged in white water kayaking down rapids. However, should the young Chinese female in this example find that all others in her canoeing group are of a background similar to her own, she might realise that the actual 'danger' or 'challenge' being posed is comparatively small. On the other hand, if her group comprises young males with a lot of white water experience, she may have concerns for her well-being.

In this example the composition of the group is being interpreted as symbolic of the degree of adventure and risk being potentially entailed. It is a form of knowledge — a form of communication about risk. Figure 4.2 incorporates 'knowledge' of risk, but such knowledge as in all cases of tourism is an expectation of an experience prior to consumption of the product. Espiner (2001) has studied the communication of hazards in natural settings (that is the Glacial fields of New Zealand's Southern Alps) and concludes that visitors had weak understandings of the risks involved. To what extent such findings can be generalised is uncertain, but direct communication with adventure product operators would lead this author to the conclusion that in many instances the sources of risk are indeed poorly understood by many visitors. Given the risk of undertaking activities beyond their levels of skill, the role of the guide/instructor becomes paramount in instilling confidence, shifting the skill level of the client upwards by carefully constructing and controlling the experience as far as is possible, and, if need be, in denying participation.

Secondly, an assessment of the value of the risk entails a comparison with the benefit to be derived from undertaking the 'risky' action. The benefit, it is argued, is primarily derived from the feelings associated with a peak experience, but it has to be recognised that such benefits are not the sole ones. Carr's (2001) work with mountain climbers reveals that even within this group engaged upon what might be termed 'hard adventure', the motives of enjoying the outdoors, experiencing new environments and appreciating the beauty of nature were among the most important motives. As noted, this is consistent with the concept of wilderness as a transcendental experience (Arnold *et al.* 1998). This desire for peak experience is both a reason for doing something (a 'pull' factor in conventional tourism push-pull theory) and represents an escape from the normal patterns of work.

Figure 4.2 also highlights the role of a sense of achievement — and arguably it is this sense of achievement that represents the enduring aspects of the adventure long after the exhilaration of 'the peak experience' has passed. The 'peak experience' might be said to be the feelings of immediate achievement of that moment, of a loss of relationship with the stream of one's life as it becomes consumed with the matter of the moment — it is an identification of self-consciousness with the action of the moment. It is intense. It is the surmounting of a challenge. The subsequent sense of achievement becomes a means of reclaiming the feelings experienced at the time, but the recall is also examined in terms of what it means for that person as a part of their individuality, of their personality. It is the comfort derived from the past of what the person once did as they face subsequent challenges. In this sense the 'sense of achievement' is a carrying forward into the future that which was once done in the past under challenging conditions.

Given such feelings, there is the benefit to be derived from revaluing oneself as having participated in and of meeting a challenge posed by the adventure. Additionally the retelling of the story may add to one's perceived status among one's peers.

In this analysis of risk, the gap between perceived risk and desired benefit is a relational gap. The greater the risk, it can be argued, the greater may be the benefit — but equally too may be the potential 'dis-benefit'. From one perspective it could be argued that Figure 4.2 represents a re-working of Csiksentmihayli's (1975) concept of flow. For Csiksentmihayli 'flow' occurs at the congruence of levels of challenge and skill, whereas it could be argued that within Figure 4.2 a congruence of risk and benefit represents a point of decision. If the benefits outweigh the perceived risk, then the activity will be engaged upon. However, second thoughts would indicate that the relationship is not that simple. Moderate benefits, however, measured, may be insufficient to obtain a decision to undertake an activity where the risks are perceived to be low. However, paradoxically the model also indicates that an activity may be undertaken where risks are comparatively low if the benefits are assumed to be great and possibly the risks thought to be more than in practice actually exist.

As has been argued, risk is associated with likelihood. In the example of sea kayaking the level of risk during summer may be low, but the benefits may be high, not so much in terms of 'adventure' but in terms of unaccustomed physical effort to enjoy solitude and nature which have also been shown to be benefits derived from 'adventure products'. However, the potential risk associated with changes in weather might also be present. Thus a distinction may be created between actual and potential threat to well-being.

It is here that the role of being on holiday might be said to be important. For many tourists the type of adventure product engaged upon is not the premeditated and planned for activity like mountain climbing that is described by Carr (2001). Rather it is the taking of opportunity to undertake an activity not normally available, whether it is white water rafting or, as described by Espiner (2001), a walk to a glacier face. If the characteristics of holidaying are those of relaxation, of seeking agreeable experience, and while wanting a challenge, having an expectation of an achievable 'high', then arguably the holidaymaker is in a state of a lack of preparedness for risk. Indeed, the very lack of preparation may increase the risk. Hence the cases reported in countries like New Zealand and Australia of walkers and trampers getting lost and being at risk of heat, cold or lack of water through not undertaking simple precautions of having suitable clothing, sufficient food and water.

The diminution of risk therefore seems to lie in the communication of what is required for safety. The issue may be that for many tourists adventure tourism is not that which was defined by Ewert (1989). As cited at the start of this chapter, for Ewert, adventure tourism is about the uncertainty of outcome. The commercial reality of many of today's adventure tourism products is that tourists would be reluctant to book product where outcomes were uncertain. Rather, what is required is the certainty of achievement — of being able to achieve within their limits a feeling of surmounting challenge. In consequence the commercialism of the adventure leads to the construction of an experience where both company and client play a game of shared disbelief. The company promises adventure and its accompanying risks while trying to ensure that risks for any given level of ability are minimal and that satisfaction will be the outcome. Nature is 'coerced' into being a resource to satisfy the wants of escape, achievement, challenge and physical release required by the tourist. The tourist on the other hand wants to believe in that presence of challenge, but equally entrusts the company to deliver the means toward achievement. Yet, equally the tourist does not want to know of the means by which the company seeks to secure satisfaction as that demeans the sense of achievement.

However, within this scenario of shared games, nature does not always play the role allocated to it, and thus the potentiality for danger is always present. It is that likelihood, however unreal it is for the most part, that provides the justification for the sense of adventure. It is argued, therefore, that the relationships described in Figure 4.2 are capable of capturing a range of realities that relate to risk taking and adventure tourism. The diagram highlights the non-linear relationships of the gap between benefit and threat, and further identifies variables that need to be considered when planning risk and adventure in a commercial setting where uncertainty is to be minimised and the opportunity for achievement is to be maximised. The paradox therefore, is that the product comes to mirror the very commercialisation processes from which the tourist sought to escape.

References

Arnold, E. J., Price, L. L., & Tierney, P. (1998). Communicative staging of the wilderness servicescape. *Service Industries Journal, 18* (3), 90–115.

Bentley, T. A., Page, S. J., & Laird, I. S. (2000). Safety in New Zealand's adventure tourism industry: The client accident experience of adventure tourism operators. *Journal of Travel Medicine, 7* (5), 239–245.

Bernstein, P. L. (1996). *Against the gods: The remarkable story of risk.* New York: John Wiley & Sons.

Carnelley, K., & Ruscher, J. B. (2000). Adult attachment and exploratory behaviour in leisure. *Journal of Social Behaviour and Personality, 15* (2), 153–165.

Carr, A. (2001). Alpine adventurers in the Pacific rim: The motivations and experiences of guided mountaineering clients in New Zealand's southern alps. *Pacific Tourism Review — An Interdisciplinary Journal, 4* (4), 161–169.

Chang, E., Dixon, K., & Hancock, K. (2001). Factors associated with risk-taking behaviour in Western Sydney's young people. *Youth Studies Australia, 20* (4), 20–25.

Csikszentimihalyi, M. (1975). *Beyond boredom and anxiety.* San Francisco: Jossey-Bass.

Elms, D. (1998). Risk management: General issues. In: D. Elms (Ed.), *Owning the future: Integrated risk management in practice* (pp. 43–56). Christchuch, New Zealand: Centre for Advanced Engineering, University of Canterbury.

Espiner, S. R. (2001). Visitor perception of natural hazards at New Zealand Tourism attractions. *Pacific Tourism Review — An Interdisciplinary Journal, 4* (4), 179–200.

Ewert, A. (1989). *Outdoor adventure pursuits: Foundations, models and theories.* Columbus, Ohio: Publishing Horizons.

Farthofer, A., & Brandstaetter, H. (2001). Extraversion and optimal level of arousal in high-risk work. In: H. Brandstaetter, & A. Eliasz (Eds), *Persons, situations and emotions: An ecological approach. series in affective science* (pp. 133–146). New York: Oxford University Press.

Fung, T. T., Hu, F. B., Chu, N. F., Spiegelman, D., Tofler, G. H., Willett, W. C., & Rimm, E. B. (2000). Leisure-time physical activity, television watching, and plasma biomarkers of obesity and cardiovascular disease risk. *American Journal of Epidemiology, 152* (12), 1171–1178.

Gibson, H. J. (1996). Thrill seeking vacations: A life span perspective. *Loisir et Société, 19* (2), 439–458.

Giddens, A. (1998). Risk society: The context of British politics. In: J. Franklin (Ed.), *The politics of risk society* (pp. 23–34). Oxford: Polity Press.

Henderson, K., & Ainsworth, B. E. (2001). Researching leisure and physical activity with women of color: Issues and emerging questions. *Leisure Sciences, 23* (1), 21–34.

Howard, R., Yan, T. S., Ling, L. H., & Min, T. S. (2002). Risk taking and metamotivational state. *Personality and Individual Differences, 32* (1), 155–165.

Keey, R. (1998). Australia/New Zealand risk management standard. In: D. Elms (Ed.), *Owning the future: Integrated risk management in practice* (pp. 91–97). Christchuch, New Zealand: Centre for Advanced Engineering, University of Canterbury.

Lash, S., Szerszynski, B., & Wynne, B. (1996). Introduction: Ecology, realism and the social sciences. In: S. Lash, B. Szerszynski, & B. Wynne (Eds), *Risk, environment and modernity: towards a new ecology* (pp. 1–26). London: Sage.

Lindstroem, M., & Sundquist, J. (2001). Immigration and leisure time physical inactivity: A population-based study. *Ethnicity and Health, 6* (2), 77–85.

Maslow, A. H. (1954). *Motivation and personality.* New York: Harper and Row.

McGuiggan, R. L. (1999). The Myers-Briggs type indicator and leisure attribute preference. In: A. G. Woodside, G. I. Crouch, J. A. Mazanec, M. Oppermann, & M.Y Sakai (Eds), *Consumer psychology of tourism, hospitality and leisure* (pp. 245–267). Sydney: School of Marketing, University of Technology Sydney.

Pondé, M. P., & Santana, V. S. (2000). Participation in leisure activities: Is it a protective factor for women's mental health? *Journal of Leisure Research, 32* (4), 457–472.

Ryan, C. (1997). Rafting in the Rangitikei, New Zealand — An example of adventure holidays. In: D. Getz, & S. J. Page (Eds), *The business of rural tourism — international perspectives* (pp. 162–190). London: International Thomson Business Press.

Ryan, C., & Ruthe, J (2002). Language and perception of a New Zealand adventure location. In: C. Pforr (Ed.), *12th international research conference of the council of Australian universities in tourism and hospitality education, Freemantle.* Proceedings of the Conference. Available on CD-Rom from CAUTHE.

Walle, A. H. (1997). Pursuing risk or insight: Marketing adventures. *Annals of Tourism Research, 24* (2), 265–282.

Wang, W., Zhum, S. Z., Pan, L. C., Hu, A. H., & Wang, Y. H. (2001). Mismatch negativity and personality traits in chronic primary insomniacs. *Functional Neurology, Special Issue — New Trends in Adaptive and Behavioural Disorders, 16* (1), 3–10.

Chapter 5

Current Status and Future Directions in the Adventure Tourism Industry

Alan Ewert and Lynn Jamieson

"Thousands of tired, nerve-shaken, over-civilized people are beginning to find out that going to the mountains is going home; that wilderness is a necessity; and that mountain parks and reservations are useful not only as fountains of timber and irrigating rivers, but as fountains of life."

(John Muir, 1901)

Introduction

This year, it was not unusual to find thousands of people scaling tall and forbidding mountains, rafting down rivers frothy with white-water, trekking in vast wilderness areas, SCUBA-diving on beautiful coral reefs and exploring the "dark wilderness" of numerous underground caverns and caves. Indeed, the year 2001 saw 183 people reach the top of Mount Everest. This total was the highest number of successful climbers in any one year, and raised the grand total of summiteers to 1,114.

Despite the broad diversity of venues and experiences, one commonality that links all of the previously mentioned activities and participants is the rubric known as *Adventure Tourism*. This chapter provides an overview and synopsis of some of the major issues and trends currently being experienced in the Adventure Tourism field. It should be noted that Adventure Tourism as an academic line of inquiry is a relatively recent entry and as such, much of the data, particularly those related to participation are often suspect and prone to multiple interpretations. Despite this shortcoming, one trend seems to be fairly consistent — namely, the continued growth in popularity, spectrum of activities, and anticipated goals in the use of these types of activities (Molitor 2000).

While acting as a catalyst for attaining high quality leisure and recreation experiences for a growing segment of the population, the growth in Adventure Tourism has also precipitated a variety of issues and challenges, both in the impacts to the natural

Managing Tourist Health and Safety in the New Millennium
Copyright © 2003 by Elsevier Science Ltd.
All rights of reproduction in any form reserved.
ISBN: 0-08-044000-2

resource base in which these activities take place, as well as the recreational experiences of users. This chapter explores some of the salient variables influencing participation in adventure recreation, and concludes with some implications for future participant and managerial actions.

The Adventure Tourism Experience

There exists now a wide spectrum of activities that fall under the caption of Adventure Tourism. Several of these activities including rock climbing, white-water rafting, mountaineering, caving, mountain biking, and canyoneering. Participation in these activities involves a wide variety of settings, skill requirements and equipment, both from the perspective of the individual and the group. Moreover, a great diversity is evolving concerning the types of programs and organizations that offer Adventure Tourism. These involve for-profit, not-for-profit, non-governmental organizations (NGO's), and similar organizations. For the purpose of this chapter, we are defining Adventure Tourism as:

> A self-initiated recreational activity, typically involving a travel and overnight stay component, that usually involves a close interaction with the natural environment, structurally contains elements of perceived or real risk and danger, and has an uncertain outcome that can be influenced by the participant and/or circumstance (Ewert 2000).

As such, this definition encompasses five major characteristics:

(a) a travel and overnight stay component;
(b) involvement with a natural environment;
(c) elements of risk and danger;
(d) uncertain outcome; and
(e) influenced by the participant or circumstance.

In addition, the adventure tourism activity often includes a small group of participants, visiting an exotic and/or remote location, often under somewhat primitive conditions, and utilizing the services of a guide, leader, and outfitter.

Whatever the activity under scrutiny, there are a number of ways in which the Adventure Tourism experience can be conceptualised and conducted. Figure 5.1 provides one such example. Of particular importance is the fact that many of these activities are amenable to diversification in how the experience is conducted, the type of client that would be attracted or interested, and the anticipated outcomes from that experience. For example, the *Adventure Tourist* on a cruise to the Antarctic may have substantially different reasons for participating and expectations than the *Adventure Traveller* on a commercial cruise to an exotic SCUBA location.

Moreover, the adventure experience can vary along a number of dimensions, including type of travel, group membership, and/or amount and spectrum of risk. This variance is also illustrated in Figure 5.1. As depicted, the participant can engage in the

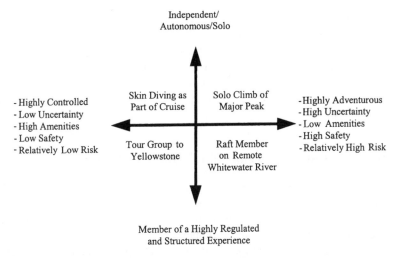

Figure 5.1: The adventure tourism experience.

adventure activity along several dimensions. In part, the location on these different dimensions (e.g. remote wilderness trip by oneself versus a trip to Cancun as a member of a cruise trip) points to the differing levels and types of risk and danger expected to be encountered (Bentley & Page 2001).

In a similar fashion, the Adventure Travel industry uses potential risk as a parameter in categorizing adventure activities into Hard or Soft Activities. Table 5.1 provides a representation of this categorization approach.

Despite the fact that exceptions can be found in each example, the above categorization system clearly uses a risk-potential factor as one way of distinguishing between these activities. Other ways to provide a distinguishing mechanism would be to include needed technical skills, perceived difficulty of the activity, and overall physical and emotional demands placed on the participant. For example, mountaineering generally involves greater risk, demands a higher level of specialized skills, and places the participant in more challenging environments than does biking as an activity.

Table 5.1: Hard and soft adventure activities.

Hard activities	Soft activities
Rock and mountain climbing	Camping
Snorkeling/SCUBA diving	Biking
Caving	Flat-water canoeing
Whitewater boating	Photo safaris
Wilderness backpacking	Day-hiking

Once again, while exceptions can be found in each example, classification systems, such as the previous example, can provide for some level of understanding as to the nature and demands of a particular activity. Given the wide variety between, and within specific Adventure Tourism activities, it is important to be able to distinguish between one type of activity and another on more variables than just location. At the same time, location does provide a useful dimension in discussing and researching the construct of Adventure Tourism.

Adventure Tourism and Traditional Tourism Activities

Adventure Tourism is a subset of the tourism industry and its growth has paralleled that of the general industry in many ways. Historically, any form of travel was considered dangerous, and only for war and business purposes. Thus, travel developed as a result of the invention of money, and recreational travellers were willing to put up with tremendous inconveniences (Crossley *et al.* 2001). The great empires including Egypt, Assyria, and Babylonia improved roads for military purposes, which then gave rise to recreational tourism, again for only the strongest of recreational consumers. It was not until the 19th century that Thomas Cook organized tours that fall into the Adventure Tourism category. Known to this day as a "Cook's Tour," these adventures included areas of natural beauty such as Switzerland's Alps, the Nile, Mount Everest, India, and Yellowstone Park. In general, the pattern of Adventure Tourism was dependent, as with all tourist opportunities, on the development of highways and other infrastructure areas. Thus, the growth of Adventure Tourism was in direct proportion to the accessibility of the areas to be visited, as well as the conditioning of the tourist.

Technological advances also permitted greater and greater travel into remote areas. Ease of transportation, weatherproof gear, durability of equipment and increased tour guidance allowed even the most uninitiated to enjoy a part of the remote outdoors in new ways. As a result, more and more people began pursuing "the great getaway" to enjoy periods of natural wonder, isolation, and adventure.

Adventure Tourism follows the same historical background as other tourism opportunities and its popularity has been dependent on several components: the availability of free time and discretionary income; advances in technology; economic conditions, availability of infrastructure such as roads, resources, and accommodations; and willingness to pay characteristics of the consumer.

Components of Adventure Tourism

Adventure Tourism is dependent upon the same three components as tourism in general: the market, the destination, and the linkage. Blank (1989) noted these three components as essential parts of the tourism system within communities. The identifiable market "consists of all people who come as tourists to a given community" (Blank 1989: 8). Identifying these tourists and categorizing their leisure interests allows for tourism planning and marketing efforts. Further, destination management factors allow for an effective infrastructure to be available for those who access mass tourism areas and those who access more remote and singular areas. Finally, linkage of information between the market and destination managers is considered an important component in any tourism effort.

Adventure Tourism is unique in several ways due to the specialized nature of the destination experience and the risk/skill factors that come with the tourist. In looking at the three components of traditional tourism activities, and comparing them with Adventure Tourism activities, there are several distinctions that are noteworthy:

(1) Adventure Tourism activities require a much greater amount of preparation on the part of the consumer than traditional activities. Very often, they require physical training, certifications, trip/excursion planning, staff with exceptional expertise, and a high level of risk;

(2) The above factors then require a much larger investment of funds, both on the part of the consumer and also the trip management organization. For example, the book *Into Thin Air* (Krakauer 1997), reveals that the average investment for an assault on Mount Everest was $65,000. Therefore, the average tourist is often a wealthy, accomplished person with time and monetary resources to afford an Adventure Tourism experience;

(3) Adventure Tourism activities often require planning to enter remote areas not usually used by the average tourist, and the entry requirements may involve waiting lists and other barriers to planning. Wilderness permits are restricted to minimum numbers in pristine areas, and often adventure excursionists have to wait for a permit to enter;

(4) Adventure Tourism business ventures often rely on seasonality and weather cycles, which impact greatly on profitability. For example, in drought areas, rivers may become inaccessible, or in ski areas, snow may not be adequate until well into the season. Such conditions will threaten a business, and as such, many businesses fail in the first year, or even after many years, the cyclic factors continue to erode profits and discourage entrepreneurs;

(5) Managing risk is a much more challenging task in Adventure Tourism activities. Quite often, professional liability concerns outweigh desire to become a professional tour planner and guide. This issue may be a reason for closing a business. Therefore, tour companies are often undergoing changes of management or becoming extinct, even if the demand for such services rises;

(6) The time it takes to plan, implement, and complete an Adventure Tourism activity may be prohibitive to most individuals who have jobs, school, family or other commitments. Marketing challenges are paramount in order to secure a market share of individuals, and some demand estimates for adventure oriented activities has declined or reached a plateau, according to Roper Polls (2000);

(7) Finally, the availability of trained and competent personnel continues to be a problem due in part to the time it takes for one to become certified in the Adventure Tourism areas of responsibility. Consequently, many tours involve those with minimal training placing tourists at the mercy of potential incompetence.

Nonetheless, the Adventure Tourism market continues to grow when taking into consideration those activities now available in simulated environments. For example, one may gain training and experience in climbing on indoor climbing walls without ever leaving town.

Adventure Tourism and the Relationship to Traditional Tourism Activities

Adventure Tourism has a dependent relationship to traditional tourist activities in that many destinations include a wide variety of venues for entertainment, sport, and pursuit of natural wonders. Most tourism literature uses the natural beauty of a destination as a pull factor or draw for tourists. All types of tourists descend upon an area, and the adventure tourist separates from the mass tourist to pursue those activities most suited to solitude, achievement, challenge, and risk. At the same time, the adventure tourist may also access other venues in the shoulder periods of the vacation: such as restaurants, attractions and amusements, lodging, sports and performances. It is least likely that the traditional tourist will access the level of adventure that the adventure tourist experiences; however, a recent *Nature Company Guide* highlights many features of "nature travel" that might appeal to both the mass and adventure tourist. This guide features natural areas of beauty that can be enjoyed by the amateur outdoor enthusiast as well as the expert. It also features information about preparing for such a vacation through conditioning, proper clothing and equipment, and logistics of the trip. Each venue featured in the book is described and includes restrictions, hours, fees, access for all including those with disabilities and mapping needs. The main feature of this book is a guide for all in order to allow for sustainable tourism (Wallace *et al.* 1995).

For the adventure tourist, as well as the traditional tourist, logistics of planning are important factors in an enjoyable experience, and many travellers fall prey to inconveniences such as food-borne illness, climate changes that affect comfort and enjoyment, exposure to new cultures that may be challenging, and reverses in plans that could interrupt or terminate a trip. Preparation of items such as proper clothing, first aid equipment, contact information, travel itineraries and other similar logistics are much in common regardless of the end goal. The destination must be reached and a schedule needs to be followed, money needs to support the efforts, safety and security are paramount to enjoyment, and the ultimate experience must be positive. Interruption in any of these areas can cause great distress and disappointment regardless of the "level" of adventure.

What *does* differentiate the adventure tourist from the traditional tourist is, primarily, the level of expertise needed to accomplish the goal. Thus training, accuracy in planning, and sensitivity to environmental issues are magnified in one's effort to assume risk and challenge physical endurance. The following areas are concerned with the adventure tourist, and these areas must be addressed in order to secure a sustainable as well as an enjoyable venture.

(1) Environmental awareness — Adventure Tourism requirements include one's ability to "leave no trace", and while this concept is supported in the mass tourist natural environments as well, the adventure tourist must become schooled in many areas: packing in and packing out everything brought in, maintaining hydration, handling

water quality issues, dealing with indigenous animals, handling unique medical conditions brought on by exertion, altitude, heat and cold weather conditions, storms, floods and other natural occurrences, and maintaining a natural environment;

(2) Development of expertise — Adventure tourist ability ranges from amateur to expert; however, all tourists in this "expert" category may encounter situations that require expertise beyond training. The ability to recognize the need for training in emergency preparedness, be cognizant of potential dangers to oneself and others, and be able to react properly to dangerous conditions are requirements for a successful Adventure Tourism guide. As noted in many well-publicized Mount Everest assaults, even the top experts in climbing could not overcome the disastrous weather conditions that lead to several deaths. Even the most simplistic Adventure Tourism activities could turn out to be deadly if certain conditions exist;

(3) Risk awareness and risk management are important to a successful Adventure Tourism activity, and both the tourist and the leaders of such excursions must be able to recognize hazards that cause injury and control for risk through tight supervision, inspection of equipment, practice, and advanced training.

In short, the Adventure Tourism experience is enhanced through the totality of preparation that is accomplished through pre-trip planning, actual experience, and recollection. The total experience can be one of the greatest accomplishments one makes, and the singularity of the experience can be life changing if these tenets are met.

Marketing for the Adventure Tourism Clientele

Moscardo (1999) notes the importance of the communication message that is conveyed to visitors. This message is often unclear, and the result tends to become a criticism of the tourist rather than the issue of failing to provide an accurate expectation for the tourist. Of greatest importance in Adventure Tourism is the degree to which clientele may be made aware of the message of sustainable tourism.

An example of two different ways that tour bus operators handled this topic is provided to show the impact of a failure to communicate the idea of sustainability. In the Wawona area of Yosemite, two tour busses approached a picnic area to stop for lunch. At this location there were picnic tables, and areas for the guests to enjoy a lunch. There were garbage cans for disposal of waste and beautiful viewing areas. One bus unloaded its passengers who ate a sack lunch, and disposed of this lunch in the garbage cans provided. They then left the area to continue on their tour. A second bus unloaded its passengers with Styrofoam box lunches. These boxes were too large to be disposed of in the "animal-proof" garbage cans so people placed them on top of the cans. Coyotes appeared and very aggressively approached the visitors who fed them scraps from lunches. These coyotes also ate food left over from the garbage area. It was obvious that the coyotes arrived with each bus.

In the true case example above, the message of sustainability was only delivered to one of the tours. Coyotes were not attracted to this group as readily as the one that used the Styrofoam containers. The message coming from the tour company could have been simple and effective; however, not all tour companies are consistent with the message.

The quality of the tourist experience is critical in providing continued enjoyment of a natural area. Tourists, according to Moscardo (1999: 7), must be made aware of the "biological diversity and maintain ecological systems". This goal requires a number of messages that can provide ways to sustain the environment such as:

(1) Operating within the limits of the resource by limiting impacts;
(2) Providing the visitor with an understanding of the environmental characteristics that sustain the experience;
(3) Vigilant adherence to capacity issues;
(4) Educating the client about activities appropriate to the region.

This effort, according to Moscardo, creates visitors who are "mindful" of the needs for sustainability (i.e. the quality of life of the environment).

This education message begins with the type of message relayed to the visitor through tourism collateral materials. The type of visitor behaviour exhibited at the destination then reveals the effectiveness of this message. This message ideally is a coordination of those who market for a destination, such as the mass tourist interests, the environmental interests, and the business interests. The effectiveness of this message may alter one's decision to visit, one's preparation for the visit, and one's behaviour during the visit. If such a message is also coordinated with messages during the visit, which might include orientations, guide lectures, brochures, signs, and enforcement actions, the visitor will most likely adhere to the messages. Masberg & Jamieson (1999) found that these messages often did not relay the true nature of sustainability of the park or natural area, and as such, tourism collateral materials often "mis-educated" the prospective visitor. Moscardo (1999: 27) identified how an effective message consists of the following:

(1) Variety/change;
(2) Use of multi-sensory media;
(3) Novelty/Conflict/Surprise;
(4) Use of questions;
(5) Visitor control/Interactive exhibits;
(6) Connections to visitors;
(7) Good physical orientation.

The above factors identify greater learning on the part of the visit at the site. The linkage for collateral materials is to carry this same type of message to the visitor prior to the visit.

For Adventure Tourism, the environmental message starts at the advertisement phase and carries through to visiting the location and enjoying a satisfying, tourism-related experience. A prospective tourist plans and prepares with the availability of accurate information: collateral materials prior to the visit, accurate logistical tools such as mapping, specific information about the sustainability and risks at the destination, accurate information about support mechanisms that may be drawn upon in an emergency or to improve the experience, and other factors that create an overall positive experience. Wallace *et al.* (1995: 11), note the value of this part of the visitor educational process through the following statement:

> Our parks, reserves, and wildlife refuges offer easy access to the wonders of the natural world, and a peaceful escape from the city. These still-wild places and the creatures that inhabit them are also an important part of our heritage. From the lush forests of the Northeast to the spectacular canyons of the Southwest, North America's landscapes and wildlife are among her greatest assets. By safeguarding places like the Florida Everglades, the Alaskan wilds, and Oklahoma's Tall Grass Prairie, we are not only protecting unique terrain and saving such creatures as the timber wolf and the American bison from extinction, but are also preserving a singular North American experience.

Variables Influencing Participation

A number of factors now serve to influence participation in adventure recreation. These variables include demographic factors, such as gender and ethnicity, as well as technological innovations. Taken in sum, these factors will generally contribute an enhancing influence relative to participation, although there will be some exceptions to this general trend. The overall estimates of participation, regardless of influencing variables, illustrate a great deal of variance as well as commonalities. Table 5.2 illustrates the data of the 1993 State of the Industry Report by the Outdoor Recreation Coalition of America (ORCA), the 1994–1995 National Survey on Recreation and the Environment (1995), and the 1997 Travel Industry Association of America (TIA) Special Report on Adventure Travel (TIA 1998).

Table 5.2: Adventure participation rates.

Activity	Number of participants (millions)		
	ORCA study	NSRE study	TIA
Hiking	22.7	47.8	44.8
Hiking to a summit	n/a	16.6	n/a
Orienteering	n/a	04.8	n/a
Backpacking	10.4	15.2	8.0
Backpacking to a summit	n/a	6.6	n/a
Mountain climbing/rock climbing	4.11	7.0	7.4
Caving	n/a	27.9	5.7
Canoeing/kayaking/rafting	14	n/a	14.8
Snokeling/SCUBA	3.2	n/a	12.4
Mountain biking	5.0	n/a	10.8

Demographic Variables

General demographic variables that will exert significant influence on adventure recreation participation and popularity include age, gender, culture/ethnic/racial background, disposable income, and level of education. Within this context, disposable income and level of education both appear to be headed in an upward direction, and both generally lead to greater rates of participation. While the disposable income variable effect is obvious, in general the relationship between education and adventure recreation participation is strong, but correlational rather than causal. As such, education may play a moderating role by creating more awareness and availability of opportunities through programs, literature, or media-based outlets.

Gender

Higher levels of participation, with respect to gender, have historically been noted for males. This trend, however, appears to be reversing with increasingly higher levels of females now participating in a wide variety of adventure recreation activities. While parity is more the norm than the exception now, one report (TIA 1998) suggests that women are more likely to be the dominant participant group for Soft Adventures while men are slightly more numerous in the Hard Adventures. As a result, while there still exist participation rate differences as a function of gender, those differences tend to be dissipated when the data are in aggregate form.

Culture/Ethnicity

Little information is currently available regarding the relationship between adventure recreation and culture/ethnicity/racial variables. Unquestionably, minority group members are not typically seen participating in adventure recreation to the same extent that other groups are, such as Afro-Americans or White Caucasians. In a similar fashion, it would seem reasonable that any real or perceived constraints to participation in adventure recreation activities would be concomitant to those experiences which take place in the more general outdoor recreation scene (see Gramann *et al.* 1993). The cause of these differences is subject to debate and is discussed in Manning (1999: 35–42). Despite this lack of theoretical understanding, it would seem reasonable to expect a greater level of participation among an increasing number of cultural/ethnic/racial groups in the future. To date, however, the extent of this growth is not known, nor is the relationship between culture and risk-taking fully understood.

In one sense, the effect of age is self-evident. That is, the older a person becomes, the less likely they are to continue participation in the more risky activities in Adventure Tourism. It should be noted, however, that participation does not entirely cease with advancing age, rather it changes to less physically demanding tasks. Conversely, Sugerman (2001) reported that reduced participation of older adults in activities such as

Adventure Tourism is more a function of lack of information rather than physical constraints. Despite these findings, the data appear fairly consistent in suggesting that most participation in adventure recreation occurs between the 25 to 44 year-old period (ORCA 1993; TIA 1998).

It remains to be seen, however, how the effects of age, as well as other factors such as quality of participation, are offset by the advances in technology. The following section discusses the role that technology will play in the Adventure Tourism setting.

Technology

It is increasingly clear that technology will continue to play an important role in the development of Adventure Tourism activities. Ewert *et al.* (2000) suggest that technology essentially takes three forms:

(a) equipment and material;
(b) education and information; and
(c) staff training.

More specifically, these three forms impact the Adventure Tourism experience through five vectors:

(a) access and transportation;
(b) comfort;
(c) safety;
(d) communication; and
(e) information.

Taken in total, technology will serve as an enabling mechanism by making it easier for the Adventure Tourist to access information regarding potential destinations and opportunities, increasing the comfort level of the trip, and provide for more safety (or the illusion of more safety; see Ewert & Shultis 1999) while engaging in that activity.

Challenges and Issues Associated with Adventure Tourism

Not surprisingly, there are a number of challenges facing Adventure Tourism. In part, these challenges and issues can be categorized as either Tourism Industry-Wide or specific to Adventure Tourism. Tourism Industry-Wide issues include sensitivity to global economic trends, the threat of terrorism, government stability, and infrastructure factors, such as the price structure of the airline industry. Issues more specific to the Adventure Tourism industry include the following:

• Environmental impacts;
• Social/Cultural/Economic impacts;

- Regulations of industry behaviours and practices; and
- Restrictions on use

Environmental Impacts

One of the largest areas of concern regarding Adventure Tourism involves the environmental and social impacts that these activities bring to an area. From the environmental perspective, impacts include soil compaction and erosion, loss of firewood, reduction in species composition and health, air and water quality degradation, and habitat disturbance. Denniston (1995) made the point that, by the 1990s, tourism in the European Alps had become a $52-billion business with over 100 million visitor-days per year. Among the obvious impacts of this high level of activity such as soil compaction and wildlife disturbance, there have also occurred tremendous increased loadings in particulates and other pollutants such as lead, hydrocarbons, and nitrogen. In sum, the environmental impacts of activities related to Adventure Tourism are fairly well known and documented. In addition, Adventure Tourism organizations and activities also can profoundly impact the social/cultural milieu and local economies.

Social/Cultural/Economic Impacts

Adventure Tourism activities can exert significant levels of impact on the social/cultural and economic systems of local and regional communities (Yuan & Christensen 1994). These impacts generally involve the following outcomes:

- Traditional livelihoods, such as farming, are replaced with activities more associated with Adventure Tourism, such as becoming a porter or guide;
- Economic returns from Adventure Tourism operations disproportionately go to only selected members of the community, who are often employed in the company. The result is an unnatural skewing of the income distribution of the community;
- While the local environmental resources of a community are utilized or consumed, the economic profits of these businesses often go to individuals and/or organizations far-removed from the community. This process is termed "leakage" and constitutes an important area of concern in the delivery of Adventure Tourism opportunities, particularly involving remote and/or exotic locations where the local inhabitants may not have adequate legal and/or political representation (Fennell 1999). As a result, the cultural history and lifestyle of these local inhabitants can often be degraded or permutated to the whims of the visiting tourist population.

Other social/cultural/economic concerns involving Adventure Tourism activities include the impact of these activities on real estate prices, availability of affordable housing, local traffic congestion, access to recreational sites, and the community's overall sense of quality of life (McCool & Moisey 1997). Blackford (2001) also discusses the controversies resulting from tourism operations and describes land use and development issues as being particularly divisive and difficult to reconcile.

Regulations on the Adventure Tourism Industry

Visitation to the Antarctic, perhaps more than any segment of the Adventure Tourism industry, may best represent a milieu involving industry-based regulations. Bauer (2001) points out that under the Antarctic Treaty System (ATS) of 1959, Adventure Tourist operations agree to abide by the following guidelines:

- Protect Antarctic flora and fauna;
- Respect historic, scientific, and other protected areas;
- Provide for visitor safety; and
- Safeguard the pristine nature of the Antarctic.

The Antarctic is not the only location in which guidelines and recommended practices have been developed. Honey (1999) reports that there are a number of organizations that deal with Adventure Tourism venues. These include the American Society of Travel Agents (ASTA), the World Tourism Organization (WTO), and World Travel and Tourism (WTTC). For example, the ASTA has its "Ten Commandments on Eco-Tourism." From these examples, it appears clear that the tourism industry has begun to respond to the need to provide some self-regulation over the various companies and organizations that comprise the industry.

Restrictions on Use

In essence, issues involving restrictions on use revolve around the broader parameters of allowable activities, acceptable practices, facilities, and suitable landscapes. Allowable activities involve circumstances in which the managing agency prohibits specific activities. For example, climbing is prohibited in many areas. This prohibition is based on the belief that the activity damages the resource through the use of crampons and ice axes, and that it disturbs wildlife, and is overtly dangerous. The prohibition of Adventure Tourism-related activities has generally been based on four reasons:

(a) damage to the physical resources;
(b) deterioration to wildlife and vegetation resources;
(c) interference with other users and uses; and
(d) an underlying belief of the resource managers that the activities are either too dangerous and prone to litigation, or are not legitimate uses of public land or water resources.

Acceptable practices imply that while certain activities might be allowed in a particular area, there are restrictions on specific behaviours associated with that activity. Staying within the climbing example — while climbing is allowed in many locations, Adventure Tourists are often faced with a number of restrictions, such as the placement of fixed or permanent anchors, climbing only during certain times or seasons (often because of raptor nesting), or the mandated use of specific equipment (e.g. coloured chalk). The issue becomes more controversial when the climbing community views the use of bolts and fixed anchors as a necessary part of the activity, without which climbing in that location would be too difficult or dangerous. Another example of controversy involves the singling out of specific Adventure Tourism activities for special regulation while at the same time not treating other uses or users in the same manner.

Facilities involve the use of human-built structures such as raft-launching ramps, horse corrals, SCUBA-training platforms, and mountain huts that serve to facilitate the engagement of specific Adventure Tourism activities. While often antithetical in officially designated Wilderness settings, the construction of specific facilities can greatly enhance the engagement of these types of activities. Historically, the development or prohibition of facilities was one technique for controlling or discouraging use.

In many respects, all three of the previously mentioned restriction parameters (allowable activities, acceptable practices, and facilities) are linked to the underlying belief of the resource manager concerning the value and legitimacy of the specific Adventure Tourism activity. This belief structure is often based on past experience, perception of impacts, level of understanding regarding the activity, and personal philosophy regarding the use of public lands (see Manfredo 1992). There exists little data to support any hypothesis identifying what direction land managers are generally taking with respect to adventure activities, although it would appear that, like the general public, Adventure Tourism activities are increasingly seen as legitimate uses of the public lands. In a similar vein, Hammitt & Cole (1998) suggest that the role of managers is not to halt change or use, but rather to manage it within acceptable limits.

In summary, significant movement has occurred in the areas of access and opportunities for Adventure Tourism. As access and opportunity is of paramount importance to Adventure Tourism activities, it seems reasonable to expect that significant controversy will continue in certain areas and locations, and involve specific activities. Overall, however, it would appear that access and opportunities will continue to enlarge, with the general positive perception of Adventure Tourism activities held by the public.

Policy Implications for Adventure Tourism

Given the continued popularity of Adventure Tourism experiences, combined with the many controversies and impacts that can occur from these activities, it is reasonable to expect that there will be a number of questions and issues related to public policy and management. The following section describes some of those policy implications:

- Is there any long-term policy and planning development for Adventure Tourism? In many locations the answer appears to be "no." This will likely become a greater issue in the future, as a number of authors now point out, developing a tourism policy for a community, region, province, state or country can help identify long-term goals, establish priorities, develop coordinated efforts and ultimately lead to a better product (Akehurst 1992; Fennell 1999).
- What will take precedence with respect to long term policy and goal setting, recreation and tourism use or preservation? For many, this is the new environmental

issue for the new century. As Wynn (2002) indicates, many tourism activities, including those associated with Adventure Tourism, can be concerned, by some, to be antithetical to the values associated with Wilderness and "pristine recreation." In another example, Kearsley (1997) believes that the tremendous increase in wilderness tourism has contributed significantly to the degradation of many wilderness sites in New Zealand.

• Specific types of tourism, such as Adventure Tourism and Ecotourism, are often considered more "eco-friendly" and amenable to long-term sustainability than more traditional forms of tourism (Weaver 1998). However, as Burr (1995) posits, tourism activities and systems that are sustainable over long periods of time can often be elusive and difficult to maintain. Thus, simply because a tourism activity occurs within a small group context, and within a natural environment, does not automatically imply that the activity is sustainable or even "eco-friendly."

• Who will determine what organizations or companies get preferential treatment for the gaining of access and the necessary permits? In addition, if sustainability and reduced environmental and social impacts are part of the granting of preferential treatment or the granting of access rights, what agency or governmental body will assume that role? As Price (1996) indicates, tourism impacts in fragile environments can be extremely detrimental to the ecosystem, both natural and human. Furthermore, the monitoring and enforcement of tourism-based regulations in remote areas can present a host of challenges and issues. With the exception of some internationally recognized areas such as the Antarctic or Galapagos Islands, little systematic and coordinated efforts have occurred in the areas of assessment, monitoring, and enforcement.

Conclusion and Summary

This chapter has provided an overview of one of the newest and fastest growing sectors of the tourism industry — Adventure Tourism. From a management perspective, the combination of exotic and remote locations, natural and unregulated environments, and the deliberate inclusion of activities that contain elements of risk and danger, provide both an interesting and challenging milieu. While environmental and social impacts are not new to the tourism industry, they can be exacerbated in the Adventure Tourism setting due to the need for pristine environments, often intense connection with the local population and culture, and the non-traditional nature of many of the activities. In addition, the recruiting of high quality staff to both facilitate the Adventure Tourism experience and safeguard the natural and cultural environment will continue to be a challenge for companies and organizations offering Adventure Tourism activities. In conclusion, while the future looks promising for Adventure Tourism, and its popularity will continue to grow, this growth will be accompanied by continued controversies. How organizations, the industry, governmental bodies, and the public deal with these controversies, will in a large part, determine the future directions and viability of Adventure Tourism, as well as its role in the tourism industry.

References

Akehurst, G. (1992). European community tourism policy. In: P. Johnson, & B. Thomas (Eds), *Perspectives on tourism policy* (pp. 215–231). London: Mansell.

Bauer, T. G. (2001). *Tourism in the antarctic: Opportunities, constraints and future prospects.* New York: The Haworth Hospitality Press.

Bentley, T. A., & Page, S. J. (2001). Scoping the extent of adventure tourism accidents. *Annals of Tourism Research, 28* (3), 705–726.

Blackford, M. (2001). *Fragile paradise: The impact of tourism on Maui, 1959–2000.* Lawrence, KS: University Press of Kansas.

Blank, U. (1989). *The community tourism imperative.* State College, PA: Venture.

Burr, S. W. (1995). Sustainable tourism development and use: Follies, foibles, and practical approaches. In: S. F. McCool, & A. E. Watson (Eds), *Linking tourism, the environment, and sustainability* (pp. 8–13). Ogden, UT: U.S. Department of Agriculture, Forest Service, Intermountain Research Station (USDA Technical Report INT-GTR-323).

Crossley, J., Jamieson, L., & Brayley, R. (2001). *Commercial recreation and tourism: An entrepreneurial approach.* Champaign, IL: Sagamore.

Denniston, D. (1995). Sustaining mountain peoples and environments. In: L. Starke (Ed.), *State of the world, 1995* (pp. 38–50). New York: W. W. Norton and Company.

Ewert, A. W. (2000). *Outdoor adventure recreation and public land management: Current status and emerging trends.* Paper delivered at the Tenth Annual World Congress on Adventure Travel and Ecotourism, Anchorage, AK, September 11–14.

Ewert, A. W., & Shultis, J. (1999). Technology and backcountry recreation: Boon to recreation or bust for management? *Journal of Physical Education, Recreation, and Dance, 70* (8), 23–31.

Ewert, A. W., Shultis, J., & Webb, C. (2000). Outdoor recreation and technologies: A Janus-faced relationship. In: I. Schneider (Ed.), *The proceedings of the 2000 social aspects of recreation research conference* (pp. 241–252). Tempe, AZ: Arizona State University.

Fennell, D. A. (1999). *Ecotourism: An introduction.* New York: Routledge.

Gramann, J. H., Floyd, M. F., & Saenz, R. (1993). Outdoor recreation and Mexican American ethnicity: A benefits perspective. In: A. W. Ewert, D. J. Chavez, & A. W. Magill (Eds), *Culture, conflict and communication in the wildland-urban interface* (pp. 69–84). Boulder, CO: Westview Press.

Hammitt, W. E., & Cole, D. N. (1998). *Wildland recreation: Ecology and management* (2nd ed.). New York: John Wiley & Sons.

Honey, M. (1999). *Ecotourism and sustainable development: Who owns paradise?* Washington, D.C.: Island Press.

Kearsely, G. W. (1997). *Wilderness tourism: A new rush to destruction.* Paper presented at the Centre for Tourism, University of Otago, Dunedin, New Zealand, 21 May.

Krakauer, J. (1997). *Into thin air: A personal account of the Mount Everest disaster.* New York: Villard.

Manfredo, M. J. (Ed.) (1992). *Influencing human behavior: Theory and applications in recreation, tourism, and natural resources management.* Champaign, IL: Sagamore.

Manning, R. E. (1999). *Studies in outdoor recreation: Search and research for satisfaction.* Corvallis, OR: Oregon State University Press.

Molitor, G. T. T. (2000). *Here comes 2015: The onset of the leisure era.* Paper presented at the Travel Business Roundtable Annual Meeting, Crystal City, VA., May 3, 2000.

Masberg, B. A., & Jamieson, L. M. (1999). The visibility of public park and recreation facilities in tourism collateral materials: An exploratory study. *Journal of Vacation Marketing, 5* (2), 154–166.

McCool, S. F., & Moisey, R. N. (1997). Monitoring resident attitudes toward tourism. *Tourism Analysis*, *1*, 29–37.

Moscardo, G. (1999). *Making visitors mindful: Principles for creating sustainable visitor experiences through effective communication*. Champaign, IL: Sagamore.

Muir, J. (1901). *Our national parks*. Boston: Houghton, Mifflin & Co., The Riverside Press.

National Survey on Recreation and the Environment (1994–1995). Survey conducted for the *Outdoor recreation resources review commission*, Washington, D.C.

Outdoor Recreation Coalition of America (1993). *Human powered outdoor recreation: 1993 State of the industry report*. Boulder, CO: The Outdoor Recreation Coalition of America.

Price, M. F. (Ed.) (1996). *People and tourism in fragile environments*. Chichester, U.K.: John Wiley and Sons.

Roper Polls (2000). *Outdoor recreation participation in America*. New York: Roper Polls.

Sugerman, D. (2001). Motivations of older adults to participate in outdoor adventure experiences. *Journal of Adventure Education and Outdoor Learning*, *1* (2), 21–34.

Travel Industry Association of America (1998). *The adventure travel report, 1997*. Washington, D.C.: Author.

Wallace, D., Holing, D., & Methvin, S. (1995). *Nature travel*. New York: Time-Life Books.

Weaver, D. B. (1998). *Ecotourism in the less developed world*. London: CAB International.

Wynn, S. (2002). The Zambezi River: Wilderness and tourism. *International Journal of Wilderness*, *8* (1), 34–39.

Yuan, M. S., & Christensen, N. A. (1994, Spring). Wildland-influenced economic impacts of non-resident travel on portal communities: The Case of Missoula, Montana. *Journal of Travel Research*, 26–40.

Chapter 6

Managing Tourist Safety: The Experience of the Adventure Tourism Industry

Tim Bentley, Stephen J. Page and Ian Laird

Introduction

Earlier in Chapter 3, Page *et al.* argue that the available data sources on tourist injuries and accidents in New Zealand were fragmented and scattered across a number of agencies, and that primary research was needed to examine different dimensions of injuries. One area that is notoriously difficult to examine is the tourism industry's experience of accident and injury, not least because of the sensitive nature of these outcomes for commercial operators.

This chapter discusses the findings of a survey of tourist operators in New Zealand in the fast growing Adventure Tourism sector, which Bentley & Page (2001) and Bentley *et al.* (2001) highlight as a problem sector for tourist accidents and injuries. This chapter also considers the accident experiences of New Zealand Adventure Tourism operators and identifies specific high-risk activities by determining the number of serious and minor accidents/injuries involving Adventure Tourism clients. Client injury-incidence rates for specific activity sectors of the New Zealand Adventure Tourism industry are identified and the role of individuals, equipment, the environment and management risk factors in adventure tourist incidents/accidents are discussed. Consequently it is possible to consider the extent and quality of accident reporting amongst Adventure Tourism operators. This provides a basis upon which intervention strategies can be devised.

As a primary research study, it provides important baseline data to complement the research discussed by Page *et al.* in Chapter 3 and is the natural corollary of developing the knowledge base on tourist accidents and injuries in one specific sector of the tourism industry for an entire country. However, prior to discussing the survey, findings and implications for managing tourist safety, the Adventure Tourism industry in New Zealand is discussed.

Managing Tourist Health and Safety in the New Millennium
Copyright © 2003 by Elsevier Science Ltd.
All rights of reproduction in any form reserved.
ISBN: 0-08-044000-2

Adventure Tourism in New Zealand

Adventure Tourism is a rapid growth sector of the tourism industry internationally. The Adventure Tourism sector encompasses a wide range of diverse activities and is defined as commercially operated activities that involve a combination of adventure and excitement pursuits in an outdoor environment. Adventure Tourism is also often taken to include more passive activities associated with eco tourism (e.g. safaris and trekking in difficult terrain). The Canadian Tourism Commission (1995 in Fennell 1999: 51) has defined Adventure Tourism as "an outdoor leisure activity that takes place in an unusual, exotic, remote or wilderness destination, involves some form of unconventional means of transportation, and tends to be associated with low or high levels of activity". In this definition, adventure travel includes nature observation, wildlife viewing, as well as adventure pursuits on air, land and water. Adventure Tourism is often divided into soft and hard dimensions. Soft adventure activities are pursued by those interested in a perceived risk and adventure with little actual risk, whereas hard adventure activities are known by both the participant and the service provider to involve a high level of risk, a feature also observed in Chapter 5 by Ewert and Jamieson.

New Zealand is regarded as a major destination for overseas visitors wishing to participate in active Adventure Tourism activities and the Adventure Tourism industry has expanded in recent years as a major niche sector within New Zealand's tourism industry (Berno & Moore 1996). Approximately 10% of visitors to New Zealand participated in Adventure Tourism of some form, according to the 1992–1993 International Visitor Survey. In 1999, the figure had risen to 11% who participated in adventure activities, including the most popular activities of scenic flights, jet boating, white water rafting, mountain recreation and bungy jumping. New Zealand's Adventure Tourism industry is a major component of its international marketing strategy. As Page & Hall (1999) noted, commercial white water rafting, jet boating, bungy jumping, skiing activities and other Adventure Tourism activities are concentrated in the country's South Island. Queenstown provides a focal point for Adventure Tourism activities and is used extensively in the destination's branding of these types of activities. The ski industry alone has an economic impact of over NZ$43 million per annum, and until the recent problems on the North Island ski fields, the region averaged over one million visits per annum. The South Island ski fields attract skiers and snowboarders from Australia, Japan and North America while the North Island market is primarily domestic.

While an important attraction of many of these activities is excitement and high levels of perceived risk (Brannan & Condello 1992; Berno & Moore 1996), there is evidence that particular Adventure Tourism activities present a serious and actual risk to the health and safety of participants. Hall & McArthur (1991) reported that 70% of all Adventure Tourism injuries and 50% of all Adventure Tourism fatalities in Australia were directly associated with white water rafting. For this reason, a national study of Adventure Tourism operators was conducted to assess accidents from an industry perspective.

Research Methodology

An accident experience questionnaire was posted to 300 Adventure Tourism businesses operating throughout the North and South Islands of New Zealand. A stratified sample ensuring representation of operators from the full range of Adventure Tourism businesses was drawn from a sampling frame of approximately 400 identified businesses. The sampling frame was constructed from various sources, including the Adventure Tourism Council's database of Adventure Tourism operators, tourist guides and brochures, fliers advertising Adventure Tourism operations and a range of other publications. Operators representing 21 different activity sectors of the Adventure Tourism industry were surveyed. A large number of operators were included in the sample for certain activities (e.g. white water rafting, walking tours and kayaking), while under five operators were identified for other sectors (e.g. caving, quad biking and skydiving). Of the operators surveyed, 142 (47%) returned fully completed questionnaires. Respondents who completed the questionnaire on behalf of the business mainly comprised sole or joint owners/managers of Adventure Tourism businesses (89%), with the remainder (9%) being non-owner managers and employees (2%) in a sector dominated by small operators (Page *et al.* 1999).

The questionnaire was designed to provide information on operators' accident reporting behaviour, accident experience (in terms of injuries to their clients while participating in the activity they provide), and the perceived causes (risk factors) of accidents and incidents involving clients. The initial section of the questionnaire asked for details about the business, including the main form of adventure activity, type of business ownership, number of clients during the preceding 12 months, number of employees, and proportion of overseas clients. The second section of the questionnaire examined the accident reporting behaviour of Adventure Tourism businesses. Operators were asked to indicate whether they were required to notify any authority or specific body in the event of a serious accident involving clients, and if so, who? Further questions sought to determine the type of accidents/incidents that are investigated and recorded internally by operators, and the type of accident reporting system used to record accident/incident information. The final section required respondents to indicate the number of 'serious' and 'minor' client injuries recorded in their accident register for the preceding 12-month period. Where records were not available, the respondent was required to provide an estimate of the business accident experience over this period using a separate response field. Operators were also asked to select injury and accident types incurred by clients. Finally, operators were asked to list the most common individual, equipment, environment and management related causes of accidents/ incidents (i.e. risk factors) for the activity provided by their business and other businesses operating the same activity.

Results

The majority (95%) of the 142 Adventure Tourism businesses surveyed were either individually or jointly owned. Most businesses surveyed were in the small/medium-sized category, with 89% of businesses employing 10 or less full time equivalent (FTE) staff (including the owner/manager) and 35% employing just one or two FTE staff. Businesses surveyed had been in operation for less than five years in 20% of cases, between five and 10 years in 45% of cases, and over 10 years in 35% of cases. The 142 Adventure Tourism operators surveyed catered for some 516,722 clients during the preceding year (1998), with client numbers for each operation ranging from 10 to 35,000 clients for the main activity provided by the business. Approximately one-half of the clients of businesses surveyed were overseas visitors, and although this varied, the clientele was dependent upon activities and locations.

Table 6.1 illustrates the distribution of activities for Adventure Tourism businesses surveyed. Activities are categorised under three main groupings, based on the type of environment in which they are undertaken; land-based, water-based and aviation-related. It should be noted that a small number of activities (e.g. guided walking tours) might involve more than one type of environment, but take place mainly on land and are therefore classified as land-based. Land-based activities were the most commonly reported activities provided by businesses surveyed, representing 48% of operators, with guided walks (15 operators), mountain recreation (11 operators), eco tours and horse riding (10 operators each) being the largest sectors. Water-based operations constituted 42% of operators. The largest water-based sectors were kayaking (24 operators) and white water rafting (10 operators). Aviation-related activities made up 10% of the sample, with almost two-thirds of respondents in this sector being scenic flight operators (9 operators).

Overseas visitors to New Zealand represented the largest proportion of clients for businesses from the guided walks, scenic flights, cycle tours and marine encounter sectors, each of whom reported estimates of 60%–100% overseas visitor rates among their total client base. The lowest proportions of overseas visitors reportedly participated in kayaking, ballooning, quad biking and fishing.

The geographical distribution of Adventure Tourism businesses surveyed is shown in Table 6.2. Adventure Tourism operators who responded to the survey were located in the five main areas of Northland, Auckland, Central North Island (particularly Taupo), Rotorua, Marlborough, Canterbury (especially Christchurch area) and the Queenstown/ Wanaka region.

South Island Adventure Tourism operators reported the largest numbers of overseas clients, in particularly in the Queenstown/Wanaka, Mount Cook, Southland and Canterbury regions. Lowest estimates were observed for North Island regions, notably, Northland, Auckland, Wellington and the Central North Island. This division reflects the pattern of tourist activity in New Zealand and the reputation of these South Island regions as leading suppliers of Adventure Tourism experiences.

Accident Reporting Experiences

Operators were asked if their business was required to notify any authority or organisation in the case of serious accidents/injuries involving clients and, if so, who should be notified. Table 6.3 summarises operators' responses by activity sector and the

Table 6.1: Distributions of activities of operators and mean number of years in business, by activity sector.

Environment	Activity of Operator	No. of Operators	Percentage of sample (%)	Years in business mean	standard deviation
Land-based	All terrain vehicles (ATV)	5	3.5	7.6	3.5
	Adventure education	4	3	7	2.3
	Bungy jumping	5	3.5	5	4.1
	Caving	2	1	2.5	2.1
	Cycle touring	5	3.5	7.4	3.4
	Eco tours	9	6	9.3	13.0
	Guided walking	15	10	8.7	6.8
	Horse riding	10	7	12.9	6.4
	Mountain recreation	11	8	12.1	10.3
	Quad biking	3	2	2.7	1.2
Sub-total		*70*	*48*		
Water-based	Black-water rafting	3	2	12.7	2.1
	Diving	4	3	9.6	6.1
	Fishing	2	1	7	2.8
	Jet boating	5	3.5	6.2	3.3
	Kayaking	24	17	10.9	5.5
	Marine encounter (dolphins/seals)	7	5	8.4	2.9
	Wind surfing	3	2	7.4	3.9
	White-water rafting	10	7	12.8	4.6
Sub-total		*61*	*42*		
Aviation	Ballooning	3	2	7.7	1.1
	Skydiving/Parasailing	3	2	6.7	5.5
	Scenic flight	9	6	15.3	8.3
Sub-total		*15*	*10*		
Total		142	100		

(Bentley *et al.* 2001).

most commonly listed notification authorities and bodies for each sector. A positive ('yes') response was provided by 74% of operators surveyed, with the remainder of respondents replying 'no' (18%) and 'don't know' (8%). Adventure Tourism businesses whose operations are covered by government legislation (e.g. jet boating and scenic flights), and fall under the jurisdiction of government authorities such as the MSA and CAA, provided a 100% positive response rate to the question of notification of serious accidents/injuries. Indeed, operators belonging to industry associations such as the New

Table 6.2: Distribution of Adventure Tourism operators surveyed by location and activity concentrations by location.

Location	Adventure Tourism operators surveyed (n)	(%)	Adventure Tourism activity concentration by location
Northland	12	8	Diving
			Horse riding
Auckland	15	11	Kayaking
			Caving
			Surfing
Waikato	6	4	Caving
Bay of Plenty	7	5	Eco tours
Central North Island	15	11	Kayaking
			White water rafting
			Cycle touring
Rotorua	12	8	All terrain vehicle tour
Hawkes Bay/ Wairarapa/ Wanganui/ Taranaki	8	6	Kayaking
Wellington	5	4	Guided walking
North Island sub-total	*80*	*56*	
Marlborough	12	8	Kayaking
			Horse riding
West Coast	7	5	Guided walking
			Mountain recreation
Canterbury	17	12	Marine encounter
			White water rafting
			Adventure education
			Balloon flights
Queenstown/ Wanaka/Otago	19	13	Mountain recreation
			Scenic flights
			Guided walking
			Bungy jumping
Mount Cook NP	2	1	—
Southland/ Fiordland	5	4	Guided walking
South Island sub-total	*62*	*44*	
Total	142	100	

(Bentley *et al.* 2001).

Table 6.3: Accident notification by Adventure Tourism operators.

Environment	Activity of operator	Required to notify of serious accidents/ injuries?			Authorities or bodies to whom report accidents
		Yes	No	Don't know	
		(%)	(%)	(%)	
Land-based	All terrain vehicles (ATV)	60	20	20	OSH; LTSA; NZP
	Adventure education	75	0	25	OSH; MSA; NZP; NZOIA; ACC
	Bungy jumping	100	0	0	SNZ; OSH; NZP; LA
	Caving	50	0	50	ACC; OSH
	Cycle touring	20	80	0	OSH
	Eco tours	50	40	10	MSA; LTSA; NZP; DOC
	Guided walking	60	40	0	DOC; OSH; NZP
	Horse riding	50	50	0	OSH; ACC
	Mountain recreation	100	0	0	NZMGA; DOC; ACC; OSH
	Quad biking	100	0	0	OSH; DOC; NZP
Water-based	Black-water rafting	100	0	0	OSH; DOC; MSA
	Diving	100	0	0	MSA; NZUA
	Fishing	100	0	0	
	Jet boating	100	0	0	MSA
	Kayaking	54	25	21	DOC; NZP; MSA; SKOANZ; OSH; ACC
	Marine encounter (dolphins/seals)	100	0	0	MSA; DOC
	Wind surfing	80	0	20	HM; NZP
	White-water rafting	90	0	10	MSA; NZP; OSH
Aviation	Ballooning	100	0	0	CAA
	Skydiving/Parasailing	100	0	0	NZPF
	Scenic flight	100	0	0	CAA
Total		*73.9*	*17.8*	*8.2*	

(Accident Rehabilitation and Compensation Corporation)

CAA:	Civil Aviation Authority
DOC:	Department of Conservation
HM:	Harbour Master
LA:	Local Authority
LTSA:	Land Transport Safety Authority
MSA:	Maritime Safety Authority
NZMGA:	New Zealand Mountain Guides Association
NNOIA:	New Zealand Outdoor Instructors Association
OSH:	Department of Labour's Occupational Safety and Health
NZP:	New Zealand Police
NZUA:	New Zealand Underwater Association
SKOANZ:	Sea Kayakers Association of New Zealand
SNZ:	Standards New Zealand

(Bentley *et al.* 2001).

Zealand Mountain Guides Association (NZMGA) or the New Zealand Underwater Association (NZUA) (e.g. mountain recreation, diving) also indicated they were aware of their responsibilities regarding accident/injury notification. The lowest proportion of positive responses came from operations not covered by such bodies (e.g. cycle tour and guided walk operators).

Not surprisingly most businesses cited the Department of Labour's Occupational Safety and Health Service (OSH) as a body to whom they were required to report accidents/injuries involving clients. The extent to which OSH's jurisdiction extends to accidents to Adventure Tourism clients is not clear, however, although sections 15, 16 and 19 of the Health and Safety in Employment Act (1992) indicate a duty of care from employers and employees towards non-employees in the vicinity of the place of work. Personal communication with OSH suggests a small number of Adventure Tourism incidents are attended and investigated by OSH personnel, although the HSE Act was not intended to cover Adventure Tourism.

The smallest proportions of positive responses to the question of accident/injury notification came from the most established businesses surveyed. Operators in the under five years in operation group provided a positive response in 82% of cases, compared to 67% for the over 10 years group. This suggests that operators entering the market recently may be more professional in their attitudes towards safety, and may have better information regarding their responsibilities in this area than more established operators. Business size appears to be related to accident reporting behaviour in a more predictable manner, with 94% of businesses employing over 10 FTE staff, 77% of businesses with three to ten FTE staff, and 60% of businesses with just one or two FTE staff stating that they notified authorities of accidents/injuries involving clients. This suggests that Adventure Tourism businesses with greater resources are more likely to have an accident/injury notification and reporting system in place, and specific personnel responsible for health and safety related matters. This is an important feature in a country where small business operators employing less than 10 people are common-place in the tourism industry.

Operators were asked to indicate which accidents/incidents (in terms of seriousness of injury) involving clients were investigated and reported internally. Some 50% of operators reported that all incidents, including near-misses, were investigated within their business. Nearly one-third of operators investigated accidents resulting in injury only, while 6% of operators only investigated accidents involving serious injuries to clients. Some 16% indicated they undertook no investigation of accidents/incidents. The less well-regulated sectors of the Adventure Tourism industry (e.g. eco tours, guided walks and kayaking) most frequently reported undertaking no accident investigation. Yet this may also be explained by the fact that accidents are rare events in these relatively low-risk sectors of the industry. Smaller businesses investigated accidents/incidents with less frequency, with 28% of one to two FTE staff operations reporting that no form of investigation is undertaken in the event of an accident/incident, compared to 6% for businesses with more than 10 FTE employees.

In the case of documenting accidents/incidents involving clients, some 60% of operators indicated they produced reports for all accidents/incidents, including near-misses. Records were produced for accidents resulting in injury for 21% of cases;

serious injury only in 5% of cases, and records were never produced following accidents/incidents in 14% of cases. Highest proportions of 'no accident recording' responses were reported by less well regulated sectors such as eco tours, guided walks, horse riding and kayaking. There were no observed differences in the production of accident report records according to the length of the business' operation. However, smaller Adventure Tourism operators with one to two FTE staff more frequently indicated they did not produce accident report records (30%), compared to operators with three to ten FTE staff (6%) and the over ten FTE staff group (13%).

Recent Accident and Injury Experience

Adventure Tourism businesses were asked about their accident experience during the previous calendar year of 1998. The questionnaire asked operators to report the number of 'serious' and 'minor' client injuries recorded in their accident register for this period. Where this information was unavailable, operators were asked to estimate the number of injuries during the same period. In all, 379 client injuries were recorded by the 142 businesses surveyed. This figure represents an overall injury-incidence rate of 0.74 accidents per 1,000 clients. No client injuries were reported for the previous calendar year by 78 (55%) businesses, while 21 (15%) reported just one injury, and 15 (11%) reported two injuries. The number of recorded client injuries during the previous year ranged from 0 to 33. These statistics indicate that minor injuries are generally unreported, as it is inconceivable that more than three-quarters of Adventure Tourism operators would experience no minor accidents/injuries to their clients during a 12 month period. Just 13 of the 379 accidents recorded (3.4%) involved serious injuries, each of the 13 operators reporting one serious injury. This figurere presents an overall serious injury-incidence rate of 0.03 accidents per 1,000 clients. Importantly, some operators may not be inclined to offer information regarding their serious injury experience, it is probable that this is a fairly good indicator of the true incidence of serious client injuries among New Zealand Adventure Tourism operators, as serious injuries are most likely to be recorded in accident registers by businesses.

It is no surprise that clients of activities undertaken in water-based operations incurred some 54% of serious injuries. Indeed, three of the five diving operators responding to the survey reported a serious injury, two while white water rafting and two marine encounter operators also reported serious client injuries. Cycling is an activity which is commonly associated with serious injuries and two of the five cycle tour operators surveyed reported serious client injuries.

Client injury-incidence rates (per one million participation hours) were determined for each activity sector of the Adventure Tourism industry. Participation hours included any travel to and from the site of the activity. Client injuries per million participation hours were collapsed into four injury groups, according to relative accident experience: zero injuries; less than 100 injuries per million participation hours (low injury-incidence group); between 100 and 499 injuries per million participation hours (moderate injury-incidence group); and over 500 injuries per million participation hours (high injury-incidence group).

Clearly, aviation activities (scenic flights, ballooning and skydiving) have best reported accident/injury performance, with all but one operator reporting zero injuries during the 12 month period. This finding is not surprising, as accidents in this sector tend to be of the low frequency, high hazard variety. Land-based activities have the highest percentage of operators in the 500+ injuries per million participation hours group (18%), and the lowest in the zero injuries group (44%).

Table 6.4 shows client injury-incidence rate groupings by activity. The highest mean injury-incidence rate was found for cycle touring (7,401 injuries per million participation hours), with three of the five cycle tour operators falling in the high injury-incidence group. Most cycle tour routes are known to predominantly follow minor

Table 6.4: Injuries per million participation hours (I.M.P.H.) groups and mean client injury-incidence rates by activity sector.

Activity	Zero injuries (%)	'Low' injury gp. <100 I.M.P.H (%)	'Moderate' injury gp. 100–499 I.M.P.H. (%)	'High' injury gp. 500+ I.M.P.H. (%)	I.M.P.H. mean	I.M.P.H. standard deviation
All terrain vehicles	40	20	20	0	25	43
Adventure education	25	75	0	0	33	45
Ballooning	100	0	0	0	0	
Black water rafting	33	0	0	67	483	425
Bungy jumping	20	40	40	0	117	127
Caving	0	0	0	100	6636	8293
Cycle touring	0	0	40	60	7401	10273
Diving	25	50	25	0	125	144
Eco tours	89	11	0	0	5	17
Fishing	0	0	50	50	3164	4096
Guided walking	80	13	7	0	20	48
Horse riding	30	30	20	20	718	1344
Jet boating	60	20	20	0	33	74
Kayaking	83	13	4	0	14	62
Marine encounter	43	43	14	0	48	84
Mountain recreation	36	27	18	18	216	330
Quad biking	0	0	0	100	3096	3112
Scenic flight	89	11	0	0	7	2
Skydiving/parasailing	100	0	0	0	0	
Wind surfing	80	0	20	0	50	112
White water rafting	30	0	40	30	537	1131
Total % for accident groups	*55*	*18*	*14*	*13*		

I.M.P.H.: Injuries per million participation hours.
(Bentley *et al.* 2001).

roads, away from the major highways (see Ritchie 1998), while some involve mountainous terrain, suggesting likely locations for cycling accidents. Other high means included those for caving, fishing, quad bike, horse riding and white water rafting, all of whom had mean client injury-incidence rates of over 500 per million participation hours, and the majority of operators in the moderate or high injury-incidence groups. Lowest client injury-incidence rates were found for ballooning, eco tours, guided walks, scenic flights, kayaking, jet boating and all terrain vehicles, all of which had injury-incidence rates of below 50 per million participation hours, and the majority of operators in the zero or low injury-incidence groups. With the exception of white water rafting, high injury activity sectors of the Adventure Tourism industry are those that are not covered by Government legislation or which do not come under the jurisdiction of Government authorities or industry representative bodies.

Among the most relatively high client injury-incidence activities are pursuits that involve the possibility of falling from a moving vehicle or animal. Cycling, quad biking, horse riding and rafting all present a degree of risk of falling from a height while in motion. Falls from a height occur most commonly in these activities, with 80% of horse riding operators, 70% of quad bike operators, 60% of cycle tour operators and 40% of white water rafting operators citing falls from a height as a type of accident clients had incurred while participating in their activity. The other activity for which operators commonly reported 'falls from a height' as a type of accident incurred by clients was mountain recreation (55% of operators). One would also expect activities that involve the risk of falling from a height to commonly report high impact injuries such as limb fractures and head injuries as injuries incurred by participants. Indeed, horse riding (60% of cases), white water rafting (30%) and mountain recreation (36% of cases) operators most frequently cited limb fractures as injuries incurred by their clients. Similar patterns in the data were observed for bruising. Head injuries were reported by 40% of cycle tour operators. These findings suggest interventions to reduce the risk of injury in these sectors should focus on reducing the risk of falling (e.g. restraint, speed reduction, instruction, risk awareness, choice of terrain), and on reducing the likelihood of injury in the event of unavoidable falls (e.g. elbow and knee padding, helmets, fall injury prevention techniques).

The most common type of accident experienced by clients of Adventure Tourism businesses was reported to be slips, trips and falls on the level. Intervention measures might include the provision of footwear appropriate for the terrain and environment in which the activity is undertaken, and avoidance of particularly hazardous terrain. Slip, trip and fall accident prevention measures are particularly important for activities such as mountain recreation, guided walking, rafting and caving, where underfoot surfaces may be wet, slippery, steep or uneven, and movement on foot may take place at speed, in poor lighting, or while simultaneously attending to other aspects of the environment (commonslip, trip and fall accident risk factors — see Bentley & Haslam 1999).

A large proportion of businesses in the moderate and high injury-incidence groups were located in the Central North Island (53% of operators), Queenstown/Wanaka region (33% of operators), and Canterbury (30% of operators). All businesses located in the Bay of Plenty, Wanganui and Taranaki regions of the North Island were represented in the zero injury group.

Operators' Perceptions of Accident Risk Factors

Operators were asked to list common perceived causes of accidents/incidents involving clients of their activity. Commonly mentioned risk factors were organised into a number of interacting subsystems: latent management and organisational factors are shown to underlie individual client, equipment and environmental factors. Management and organisational factors may themselves be influenced by extra-organisational factors such as commercial pressure or a lack of experienced staff. In rafting operations, failure on the part of management/guides to ensure all clients understand instructions (management factor) may interact with language problems, and client understanding of appropriate action in the event of a capsize (client factors) in hazardous river conditions (environmental factor). Other commonly cited examples of accident factor combinations include failure to supply (or wear) appropriate footwear and the presence of slippery underfoot conditions. These perceptions are congruent with common accident types identified by operators (i.e. slip, trip and fall accidents), suggesting support for the need to reduce the risk of falls in Adventure Tourism activities. The most commonly listed accident risk factors were client failure to follow instructions, client overconfidence, exaggerating ability and experience, showing off, use of poor standard equipment and inexperience of guides.

Implications for the Management of Tourist Safety

Adventure Tourism operators' reported accident experience suggested very few serious client injuries were incurred during 1998. Serious injuries occurred most often during water-based activities (diving, white water rafting and marine encounter operations). Relatively few minor injuries were reported by operators, suggesting serious under-reporting of minor incidents and injuries in this sector of the tourism industry. Operators' responses to questions concerning their accident reporting behaviour supported this view, and may reflect a poor safety culture within certain sectors of the Adventure Tourism industry. This may be particularly true of smaller, unregulated sectors of the industry, which in a number of cases also have highest reported injury-incidence.

Highest client injury-incidence rates were observed for cycle tours, caving, quad biking, horse riding and white water rafting. It was noted these activities all involved the risk of falling from a moving vehicle oranimal. It is not surprising that these activities (with the exception of white water rafting) represent the less well regulated sectors of the Adventure Tourism industry.

Analysis of accident events and injuries sustained by clients of Adventure Tourism activities suggest injury prevention measures should specifically focus on reducing the risk of falls from heights and slip, trip and fall accidents on the level. These risks appear to be common across most sectors of the Adventure Tourism industry. Operators for accidents involving clients identified a range of individual, equipment, environmental and management risk factors. The involvement of these risk factors and their interactions in actual accidents/incidents needs to be determined through further research, and possible measures identified to reduce the risk of accidents and injuries involving clients.

The management of tourist safety in the Adventure Tourism industry cannot simply be addressed at the micro or individual business levels. This chapter raises wider policy issues for government and the extent to which existing interventions by health and safety bodies need to adopt a legislative requirement. Following a number of Adventure Tourism fatalities in New Zealand during 1995, a Ministry of Commerce Working Paper (Ministry of Commerce 1996) outlined a range of policy options available. The government preference was for minimal intervention, and industry Codes of Practice have been implemented in the most risky adventure activities. Experience in relation to Codes of Practise reported in this chapter tends to suggest they have a favourable impact on industry regulation. Given this evidence, they need to be put in place for activities that do not have them. It is also evident that, following a series of accident events, public attention and policy makers should revisit the issue of regulation verses self-regulation for the Adventure Tourism industry. There is a continued concern at government level to ensure that businesses are not burdened with a large bureaucratic red tape function associated with administration. This concern needs to weigh against that of the tourist well-being so that clients are afforded a degree of protection that avoids unnecessary risk.

References

Bentley, T. A., & Haslam, R. A. (1999). Slip, trip and fall accidents occurring during the delivery of mail. *Ergonomics, 41,* 1859–1872.

Bentley, T. A., & Page, S. J. (2001). The cost of adventure tourism accidents to the New Zealand tourism industry. *Annals of Tourism Research, 28* (3), 705–726.

Bentley, T., Page, S. J., Meyer, D., & Chalmers, D. (2001). How safe is adventure tourism in New Zealand? An exploratory analysis. *Applied Ergonomics 2001, 32* (4), 327–338.

Berno, T., & Moore, K. (1996). *The nature of the adventure tourism experience in Queenstown.* Paper Presented at the Tourism Down Under Conference, Centre for Tourism, University of Otago, 3–5 December.

Brannan, L., & Condello, C. (1992). Public perceptions of risk in recreational activities. *Journal of Applied Recreation Research, 12,* 144–157.

Fennell, D. (1999). *Ecotourism.* Routledge: London.

Hall, C. M., & McArthur, S. (1991). Commercial white water rafting in Australia. *Australian Journal of Leisure and Recreation, 1*(2), 25–30.

Ministry of Commerce (1996). *Safety management in the adventure tourism industry: Voluntary and regulatory approaches.* Wellington: Ministry of Commerce.

Page, S. J., Forer, P., & Lawton, G. (1999). Tourism and small business development: Terra incognita. *Tourism Management, 20* (4), 435–460.

Ritchie, B. (1997). *Cycle tourism in the South Island of New Zealand: Infrastructure considerations for the twenty-first century.* Paper presented at *Trails in the Third Millennium,* Cromwell, New Zealand, 2–5 December, 325–334.

Section 3

Advice and best practice

Chapter 7

Current Issues in Travel and Tourism Law

Trudie Atherton and Trevor Atherton

Introduction

The events of 11 September 2001 have profoundly and adversely affected the tourism industry.[1] When an event of this nature and magnitude occurs, and people suffer personal loss, injury or damage while travelling, the immediate cry goes up that someone must be responsible and someone must therefore pay. The question arises 'Who assumes responsibility and who pays?' This chapter examines some of the current issues in tourism law, especially those presently giving rise to litigation. It identifies the existing international and/or national regulatory frameworks and analyses the legal responsibility of the parties who collectively and individually provide tourism and travel services.

Under the World Tourism Organization/United Nations Recommendations on Tourism Statistics, tourism comprises the activities of persons travelling to and staying in places outside their usual environment for not more than one consecutive year for leisure, business and other purposes.[2] Accordingly, the legal issues are discussed in this chapter under the headings of passenger transport, traveller accommodation activities and attractions, tour operators and travel agents, and other issues.[3]

Passenger Transport

In view of the World Tourism Organization's definition of tourism, it is clear that transport plays a central role in getting tourists to and from their destinations. All

---

[1] In Middle Eastern countries like Egypt, still recovering from its own brushes with terrorists, the impact was sudden and severe and tourists stopped coming to the country.
[2] http://www.world-tourism.org/market_research/facts&figures/menu.htm
[3] For a more detailed analysis of these issues see the authors' text: Atherton, T. C. & Atherton, T. A. (1998). *Tourism, Travel and Hospitality Law.* Sydney: LBC Information Services.

tourists must travel by road, sea or air in a variety of conveniences including cars, buses, trains, ferries, boats or aeroplanes, amongst others.

By mid-2001, domestic and international air travel was increasing partly due to increased competition and lower prices in the market place. At the same time, airlines were experiencing a sharp rise in problems associated with criminal intent and passenger recklessness. Of course, this reached a climax on 11 September 2001.[4]

Hijacking

Within the past 40 years, the world has witnessed acts of terrorism including hijacking of aircraft, shipping, bus and train[5] hostage taking and the illegal use of explosives. The international community has responded to these dramatic events by negotiating for the introduction of a number of terrorism conventions, which state parties will enforce in the spirit of international cooperation. In particular, the issue of hijacking aircraft (or 'skyjacking' as it has become colloquially known) has been addressed in a number of conventions[6] including the *1971 Convention for the Suppression of Unlawful Seizure of Aircraft [Hijacking Convention]*.[7]

After the Lockerbie disaster in 1988,[8] security at airports was stepped up again and for some time it appeared that the solution to sky terrorism had been found. Of course, the events of 11 September 2001 reignited this debate when terrorists hijacked three passenger aircraft for the purpose of using these as suicide bombers to target civilian sites within the United States of America. Many thousands of lives were lost on the ground as well as all those on board the three aircraft.

The criminal liability of the hijackers is undoubted and although they died, if they had left behind assets these could and should be seized to partially compensate the victims of this dreadful crime. The families of those who died as passengers on board the planes may have additional avenues of compensation against the carriers under the existing domestic United States of America laws, or under the international regulatory regime

---

[4] See International Air Transport Association for trends: http://www.iata,org/pr/pr02febc.htm

[5] In 1994, the Khmer Rouge held up a train travelling between Phenom Phen and Sionoukville and took three tourists from Britain, France and Australia hostage, held them for ransom and later killed them.

[6] These international conventions on terrorism include the 1963 Convention on Offenses and Certain Other Acts Committed on Board Aircraft; the 1971 Convention for the Suppression of Unlawful Acts Against the Safety of Civil Aviation; the 1973 Convention on the Prevention and Punishment of Crimes Against Internationally Protected Persons, including Diplomatic Agents; the 1979 Convention Against the Taking of Hostages; and the 1997 International Convention for the Suppression of Terrorist Bombings.

[7] The Convention entered into force generally 14 October 1971 http://www.law.nyu.edu/kingsburyb/fall01/intl_law/basicdocs/HijackingConv.htm

[8] The use of plastic bombs that could be detonated while an airline was en route became possible in the mid-1980s. The most startling example of such a bombing occurred on 21 December 1988, aboard a Pan American World Airways 747 jetliner that exploded over Lockerbie, Scotland, killing all 259 persons aboard and 11 on the ground. A plastic explosive was determined to be the cause. The Federal Aviation Authority on 30 August 1989 ordered the installation of long-range bomb-detection devices, called Thermal Neutron Analysis devices, at 40 major United States and overseas airports by the end of 1991. These detectors use neutron bombardment to scan luggage for plastic explosives, and they are effective 95% of the time. http://search.ebi.eb.com/ebi/article/0,6101,31760,00.html

built upon the Warsaw Convention (1929) and its protocols.[9] However, this system is out of date and opinion is divided on whether hijacking or terrorist acts are compensatable 'accidents' under the system.[10]

Deep Vein Thrombosis (DVT)

Long haul flights have become increasingly popular in the past 50 years with the advent of faster aircraft and cheaper fares. International travel is now accessible to most people, not just the privileged few. However, one outcome of this accessibility has been the increase in the number of economy class passengers complaining of leg cramps and/or swollen feet as a result of being unable to stretch their legs properly in the confined seating spaces. In many of these cases, it now appears that these symptoms are due to the medical condition Deep Vein Thrombosis (DVT), extreme instances of which have resulted in the collapse and death of the passenger.[11]

International travellers' claims are presently governed by the provisions of the Warsaw Convention (1929) which requires proof that either there was an 'accident' or the airline breached its duty of care to passengers. In a recent case, a German court rejected a compensation claim from a passenger who developed DVT after a Lufthansa Frankfurt to Cape Town return flight in April 2000, finding that the illness was neither an 'accident' under the Warsaw Convention, nor did it arise from poor seating arrangements but was rather a pre-existing medical condition of the claimant.[12] One commentator has suggested that if this line of reasoning is adopted in other courts, then only those passengers who are fit to fly will succeed in these claims.[13]

---

[9] Unfortunately, the limit of compensation is set quite low and the evidence of proof of negligence by the airline is often difficult to obtain. http://www.forwarderlaw.com/archive/warsaw.htm The more recent Montreal Convention 1999 is not yet in force. The major feature of the new legal instrument is the concept of unlimited liability. The Montreal Convention introduces a two-tier system: the first tier includes strict liability up to 100,000 Special Drawing Rights, irrespective of a carrier's fault. The second tier is based on presumption of fault of a carrier and has no limit of liability. http://www.iata.org/legal/dep_subpage.asp?department = 1&subpage = 119

[10] For an interesting review of the United States cases see Barrett, J. (1978). Terrorism and the Airline Passenger. *New Law Journal* 128, 489–502.

[11] Melbourne-based law firm, Slater and Gordon stated it had received details of almost 2,300 alleged DVT cases, including 116 possible deaths. It said that although some cases dated back to the late 1980s, about one-half occurred in the past two years. Global calls for safeguards against DVT arose in October 2000 following the death of 28-year-old Briton Emma Christoffersen, who collapsed in the arrival hall of London's Heathrow airport after returning from Australia following the Olympic Games. http://www.cnn.com/2001/WORLD/asiapcf/auspac/02/04/australia.airlines/

[12] Frankfurt State Court 2-21 O 54/01.

[13] Tim Bryme, aircraft liability specialist at CMS Cameron McKenna.

> 'Given the difficulty of defining DVT resulting from long air flights as an accident — that is unforeseen and unexpected — under Article 17 of the Warsaw Convention, I think claimants will begin to take the broader and apparently easier direction of linking faulty products with health implications,' he says.
> http://macdonald.butterworths.co.uk/news/getarticle.asp?newsid = 1245

Sky Rage

Several factors influence and affect passenger behaviour, particularly on long haul flights. These include the in-flight service of alcohol, the banning of cigarette smoking on flights and in airport terminals prior to departure, claustrophobia and allergic reactions to certain food service items.

Some passengers behave so poorly that they must be physically restrained with handcuffs to prevent them from causing major harm to themselves, the crew or the aircraft itself. In the latter case, damage to the aircraft (e.g. by breaking windows or opening exit doors) potentially puts the lives of everyone on board the aeroplane at risk. Generally, offenders are charged with criminal offences, and the penalties range from a ban on flying for a period of time to severe fines and imprisonment in more extreme cases.[14]

Traveller Accommodation

The second key element of tourism from the concept defined at the outset of this chapter is traveller accommodation. Providers of accommodation for travellers have long had a duty of care for the health and safety of guests and a responsibility to safeguard their property. These laws are known as the Innkeepers' Doctrine and can be traced back to ancient Roman times.[15] In those days, travel was difficult and dangerous, and travellers were very much at the mercy of innkeepers and carriers who often preyed upon them along the way. To improve the safety and efficiency of travel the Romans introduced special rules for the conduct of these callings. These rules were adopted by the common law in medieval times and developed down through the cases to the present day. These rules still apply today, with local modifications, throughout most of the common law and civil law world. However, travel and accommodation have changed dramatically and these principles have been unable to adapt quickly enough to provide an adequate regulatory framework.[16]

Unlike passenger transport, there are very few international conventions regulating traveller accommodation. Attempts by the International Institute for the Unification of Private Law to establish a Draft Convention on the Hotelkeepers Contract have been

---

[14] On 15 April 2000, Roy Santamaria became the first Australian to be imprisoned for extreme in flight behaviour. He was sentenced to three years after pleading guilty to six offences, including threatening to kill a flight attendant and endangering the safety of the Melbourne-Perth Ansett flight on which he was travelling.

In a drunken state, the 31-year-old repeatedly tried to open the aircraft's cabin door as it cruised at 35,000 feet. The crew had stopped serving him alcohol two hours into the flight after he began bragging about a previous incident on a Qantas flight two years previously, for which he received a three-year good behaviour bond. http://www.smh.com.au/travel/0004/15/travel3.html

[15] Similar principles exist in Islamic Law and can be traced back into the customary laws of the Bedouin tribes in the Middle East.

[16] Statutory modifications and the increasing cost of litigation have reduced the number of cases on these matters reaching the higher law making courts. Unfortunately, this obscurity means that these special innkeepers' rights and duties are sometimes uncertain and frequently overlooked.

debated internationally since the 1930s, but have so far been unsuccessful. Self-regulation has been more successful with the International Hotel and Restaurant Association's *International Hotel Regulations* and *Code of Conduct for Hoteliers and Travel Agents*[17] providing the foundation of the international regulatory framework.

Innkeepers' Doctrine

Some of the key legal issues emerging with traveller accommodation are discussed below in the context of the special rights and duties stemming from the Innkeepers' Doctrine. The special rights and duties that apply between innkeepers and their guests are:

Rights:

- Lien over guests' property; and
- Rules of the House

Duties:

- To receive and entertain guests;
- To receive and look after guests' means of transport;
- To take reasonable care of guests' safety; and
- To safeguard guests' property.

Innkeepers' Lien

An Innkeeper's Lien is a special right to take possession of the things that a guest brings to a hotel and retain them until charges for accommodation, food, beverage, telephone and similar services provided to the guest have been paid. It does not matter whether the guest owns the things or not. Motor vehicles can be retained under this principle, except where they have been excluded by statute. The lien does not include a right to sell the things retained, except where this is provided by statute.

Rules of the House

Innkeepers have always had the right to set reasonable Rules of the House. The practice has fallen into disuse at most establishments but warrants reconsideration as part of a modern risk management strategy at a hotel, motel or especially a resort. If a guest fails to observe Rules of the House and suffers loss then this can provide some evidence that the guest was negligent, and this provides one of the innkeeper's best defences to a claim. Matters that might be covered include rules for safety, property, visitors, and for

---

[17] http://www.ih-ra.com/publications/

the use of facilities such as the carpark, swimming pool, gymnasium or business centre.

Duty to Receive and Serve Guests

The owner or lessee of a property may usually choose at will whether to permit or deny others entry to the premises. Innkeepers have a duty to accept guests provided they have space available and provided the guest is ready, willing and able to pay for it. This duty traditionally included an obligation to supply food and beverage. However, liquor cannot be demanded of an unlicensed innkeeper. With the proliferation of fast food outlets and the development of accommodation alternatives without traditional food service, arguably food can no longer be demanded at an establishment unless it has the facilities to provide it.

Duty to Receive and Look After Guests' Means of Transport

Historically innkeepers had a duty to receive carriages and stable and feed guests' horses. The transition from horse and buggies to cars and other means of transport has left some uncertainty about the present scope of this duty. English cases have held that the duty now extends to a guest's car. Although car parking is now required by planning laws many city hotels have fewer spaces than guest rooms and it is unclear what responsibility the hotelier has to a guest with a car if the car park is full. What is the responsibility of a modern hotelier if a guest turns up on a horse? The answer may well be different between a city hotel and one in the country.

Duty to Take Reasonable Care for Guests' Safety

This duty on an innkeeper is similar to that applying to any occupier of premises in general law. It is a duty to take reasonable care for the safety of guests and to ensure that the premises are as safe as reasonable care and skill can make them. Generally the accommodation provider must be at fault to be liable. Therefore, in 1975 Connie Francis recovered US$2.5 million damages for pain, suffering, mental anguish, humiliation and loss of earnings suffered when assaulted by an intruder who broke into her hotel room through a sliding door which had an inadequate lock. Ever since, the American hotel industry has been upgrading its security systems to the stage where today many United States travel agents will not book a hotel which lacks a computer card locking system for fear that they too will be sued for negligence. This duty to take reasonable care for the safety of guests is also becoming stricter in Australia. Failure to comply with relevant fire, health and safety codes is strong evidence of negligence. The High Court of Australia[18] has also recently held that a landlord remains responsible for defects in work on premises even if these were carried out by independent tradesmen.

---

[18] http://www.worldlii.org/cgi-worldlii/disp.pl/au/other/hca/transcripts/1996/B14/1.html

Duty to Safeguard Guests' Property

Under the general law, a person taking possession or custody of another's property is responsible for loss or damage to it caused by their fault. However, innkeepers are responsible for guests' property even if they are not at fault. Liability is strict and innkeepers are, in effect, insurers of guests' property. Originally designed to protect innocent guests from dishonest innkeepers, this rule came to be abused by dishonest guests who could easily claim money from innkeepers for supposedly lost property. However innocent or careful the innkeeper, the only defence was to prove that the guest was negligent and that was often a difficult task. Last century, innkeeper laws internationally were reformed to redress this problem by allowing innkeepers to limit strict liability by statutory notice. In many jurisdictions these limits have not been adjusted for inflation and are now so low they provide little real compensation.

These limits do not apply if the innkeeper was at fault and liability remains unlimited under the general law of contract, tort, workplace health and safety and consumer protection laws. Nor do they apply unless a statutory notice is provided in the form required in the particular jurisdiction. As travellers by definition are unlikely to know the local laws and as reservations are made at a distance, there is great need for international uniformity on these matters.

Alcohol Servers Liability

Provision of accommodation often involves food and beverage service and the ambit of alcohol related cases are of increasing concern to accommodation providers. It is now well established that alcohol servers and their employers will be held responsible to intoxicated patrons and even third persons injured as a result of the drunken actions of intoxicated patrons. In a bizarre case recently a patron recovered damages for injuries suffered from slipping over where another drunken patron had walked in shoes made of pork chops donned in larrikin compliance with the dress code requiring patrons to wear foot attire.[19] There is no reason in principle why this widening concept of responsibility should not also be applied to casinos and capture dealers and casino operators.

Activities and Attractions

The third key element of tourism is activities and attractions.[20] These are often the main motivation for travel and are also often the centre of hospitality operations.[21] While incredibly diverse in character, these range from totally man-made activities such as

[19] See http://www.smh.com.au/articles/2002/06/27/1023864634063.html
[20] See, for example, European Community Directive on Package Holidays Article 2. http://europa.eu.int/eur-lex/en/lif/dat/1990/en_390L0314.html
[21] Atherton, T. C., & Atherton, T. A. (1998). *Tourism, Travel and Hospitality Law*. Sydney: LBC Information Services.

theme parks, through to untouched natural attractions like wilderness areas, with many combinations of both.

With the apparent rise in accidents occurring at theme parks and during Adventure Tourism activities, it seems appropriate here to consider the regulatory framework governing both owners and operators (hosts) and visitors (guests) engaged in these types of recreational activities.

Theme Parks

At the international level, the industry is self-regulated. The mission statement for the International Association of Amusement Parks and Attractions (IAAPA) is a clear reminder of its high ideals:

> IAAPA exists to foster the highest degree of professionalism within the amusement industry; to promote the market for its goods and services; to gather and disseminate information on the industry; and to represent the interests of the industry before government — all to the end that our member companies grow and profit.[22]

National chains of operators have also developed their own codes of conduct for visitors, such as those used in the United Kingdom.[23] Usually admission to such parks is granted upon payment of a fee so that a contractual relationship exists between host and guest. Both parties are bound by the terms of that agreement, which is often displayed at the entrance to the park. In simple terms, for its part the operator warrants the safety of the rides and the visitors agree to follow instructions when using those rides.

In recent times, patrons have become more thrill seeking, demanding faster, twistier and taller roller coaster rides and free fall rides that drop passengers from heights comparable with mid-size skyscrapers. Although it appears that the technology has improved to the point where rider safety can be ensured via computers monitoring location and speed of the device and whether everyone is buckled in, it cannot guarantee that riders will ensure their own safety by following the rules.[24]

Once injury does occur, the victim usually has a choice of legal means of obtaining compensation under domestic law. These include the general law of negligence and/or contract that involve duties of care, as well as workplace health and safety laws and consumer protection legislation. Given the current attitudes of courts around the world tending to favour consumers/plaintiffs over the defendants with the deeper pockets, these cases are frequently settled out of court and usually involve the operator's public liability insurer.

---

[22] http://www.iaapa.org/

[23] See http://www.wtbonline.gov.uk/attachments/93.pdf

[24] The United States Consumer Product Safety Commission, the agency that oversees dangerous toys, says that the number of serious injuries that require treatment at an emergency room because of theme park rides increased 24% from 1974, to 9,200 in 1998.

This has led to increased nervousness on the part of many insurers to cover the operators' risk of injury through theme park rides and has resulted in extortionately high premiums. This has forced the closure of many fun park rides already in Australia, for example. Potential victims also include country fairs and school fetes.[25]

It is difficult to envisage where this might end without legislative intervention. As one commentator put it:

> Some people feel if there is an accident, someone, somewhere should pay. There needs to be an acceptance of risk in participating in a sport. It should not be a high risk. It should not be an unnecessary risk. Negligence should never be legislated out but sometimes accidents do happen.[26]

Adventure Tourism

Adventure Tourism is increasingly popular and by definition this involves increased risk. The juxtaposition of tourists and adventure result in high risk tourism activities. Some of the more extreme activities include scuba diving, ballooning, mountain climbing, cave walking/diving also known as 'pot holing', bungee jumping, abseiling, visiting countries at war, para-jumping and white water rafting amongst others. Often these activities take place in a wilderness setting.

Although tourists consciously seek out this type of experience, the adventure tour operators engaged in facilitating these activities face onerous responsibilities on two fronts. Firstly, they are liable for the tourist's safety pursuant to contract, tort, workplace health and safety and local consumer protection laws. Secondly, adventure tour operators have a responsibility for the environment in which they conduct their business to ensure that those activities are sustainable and that they neither damage nor degrade the natural environment.

Responsibility to Adventure Tourists

There is a constant tension between Adventure Tourism and risk taking. Adventure Tourism usually takes place outdoors where people and nature interact creating the adventure. Unlike theme park rides which can be controlled and monitored at all times,

---

[25] Even the experts who vouch for the rides' safety are having trouble finding anyone to insure them. Report in the Sydney Morning Herald, *How Fun is a Casualty of the Insurance Rollercoaster*, 3 July 2002.
[26] Sarah Lucas, Chief Executive of Sports Industry Australia, says the issue of personal responsibility for accidents is ignored alongside calls for law changes, better risk management and capping.
http://www.smh.com.au/news/0201/review/review5.html

there are some aspects of Adventure Tourism over which tour operators have no control, particularly the changing weather patterns.

As the weather conditions deteriorate, so too the risks to the health and safety of tourists increase. Does this mean that the tour operator should owe the tourist a higher duty of care knowing that at any time there is a real and foreseeable risk of the weather conditions changing? Or can this duty of care be discharged if the tour operator carefully explains to the tourist the greater risk of injury inherent in the activity so that the tourist fully appreciates the extent of the risk and voluntarily accepts it?

It appears that courts do indeed demand a higher duty of care from tour operators when the very nature of their business is inherently risky. For example, in July 1999, 18 tourists and three guides lost their lives in a flash flood while they were on a canyoning trip near Interlaken in central Switzerland. Six former staff members of a Swiss adventure company[27] were found guilty of negligent manslaughter for their part in the deaths. The court found that safety measures taken by the tour operator were totally inadequate, with no proper safety training for the employees. In fact, the trip to Saxet Brook should never have taken place as the approaching storm had been forecast and was visibly approaching.[28] Switzerland has since adopted a Code of Conduct for Commercial Operators of Extreme Sports and set up courses to educate guides.

In common law countries, criminal liability generally requires proof of gross negligence *beyond a reasonable doubt*. This is a very high standard of proof and rarely succeeds.[29] For civil liability, the plaintiff must show only proof of negligence *on the balance of probability*. This is a much lower standard of proof. Now that the Interlaken case has established criminal negligence on the part of the tour operator, it should be much easier for families of those who died and others who suffered injury in the tragedy to prove negligence in their civil claims for compensation.

The law also casts some responsibility upon the adventure tourist. In the general law of tort a defendant who knowingly and voluntarily accepts the risk of an activity cannot later complain. In contract law operators have traditionally sought to exclude or limit liability by special contract conditions. However, neither of these approaches provides an operator with complete protection from successful claims.

To indemnify against the possibility of civil claims, tour operators could insist that tourists engaged in these Adventure Tourism activities take out their own insurance cover against the risk of injury. However, insurance companies also want to legally transfer the risk of taking part in tourism ventures and outdoor adventure activities to customers. There is a trend now in the insurance industry to insist that customers agree not to sue if they are injured except in cases of gross negligence.[30]

---

[27] The now defunct Adventure World.

[28] Each of the three directors, Stephan Friedli, Peter Balmer and Georg Hoedle was given a five month suspended prison sentence and a $5,000 fine, and three senior guides were given shorter sentences and smaller fines. Two junior guides were found not guilty.

[29] See, for example, the 'balloon kissing' case in Alice Springs, Australia, where charges of manslaughter were laid against the balloon pilot but were ultimately dropped in favour of lesser charges. *R v. Sanby* (unreported Northern Territory Court of Criminal Appeal, CA8 of 1992).

[30] http://www.theage.com.au/business/2002/01/19/FFXSAPSU.K.WC.html

Responsibility for Environment

The World Heritage Convention (1972)[31] was an attempt by the international community to protect, present and preserve the contemporary wonders of the world and pass them on to future generations. In many respects it has been an overwhelming success with some 174 state party members as at 1 August 2002. There are 730 properties presently inscribed on the World Heritage List.[32] This is an international model for tourism in sensitive areas.

The Convention promotes listing of sites with outstanding universal cultural or natural value and supports their ongoing sustainability through protection, conservation, presentation and rehabilitation (if necessary). It recommends educational and information programs for local people to learn more about their natural and cultural heritage and encourages local involvement of host communities in the economic benefits that flow from world heritage listing. Eco tourism[33] is seen as the one tourist activity totally consistent with World Heritage values.

The World Heritage logo acts like a brand, identifying a particular site as a five star world-class attraction. It should also be a guarantee of visitor health and safety but this is not always the case.[34] In 1997, at the Valley of the Kings near Luxor in Egypt, more than 60 people died, mostly international tourists, when gunmen opened fire on them in a senseless terrorist attack. Egyptian officials quickly responded with stepped up police surveillance and security at the site, which continues.[35]

Tour Operators and Travel Agents

Tourism, travel and hospitality are intangible, composite products often sold at a distance from where they are delivered and consumed. Consequently there is a heavy reliance on others in the sale, distribution and delivery of these products.[36]

When problems occur with a package holiday, for example, the tourist has the dilemma of deciding who is legally responsible for the disappointment, personal loss or damage. Is it the travel agent, tour operator, accommodation provider, transport carrier or tour guide? None of these people would willingly accept liability.

---

[31] Convention concerning the protection of the world cultural and natural heritage. http://www.unesco.org/whc/nwhc/pages/doc/main.htm

[32] Including 563 cultural sites and 144 natural sites and 23 mixed properties.

[33] Eco tourism activities include walking, camping, boating, fishing, horse riding, mountaineering, diving and skiing.

[34] During the early 1990s tourists visiting Angkor Watt near Siam Reap in Cambodia faced a real risk of being taken hostage by the Khmer Rouge whose gunfire was often heard in the distance.

[35] 18 November 1997, Egypt's President stated that his country is still safe for visitors. http://www.cnn.com/TRAVEL/NEWS/9711/18/egypt.tourism/

[36] Atherton, T. C., & Atherton, T. A. (1998). *Tourism, Travel and Hospitality Law*. Sydney: LBC Information Services.

Package Holidays

A typical package holiday is a complex transaction. It may include:

- The client purchasing a tour through a travel agent;
- The tour being organized by a tour operator;
- The transport, accommodation and other services in the tour being operated by others;
- The hotel being managed by an international operator;
- The restaurant being operated under an international franchise;
- The risks being underwritten by an insurer; and
- All these operators engaging agents, employees and contractors.[37]

In many cases, it is difficult to define the roles and obligations of each of these intermediaries to the consumer in producing the ultimate holiday product. To provide some clarity, the Council of the European Community has produced a Directive, which in Article 2 defines a Package Holiday.

> For the purposes of this Directive:
> (1) 'Package' means the pre-arranged combination of not fewer than two of the following when sold or offered for sale at an inclusive price and when the service covers a period of more than twenty-four hours or includes overnight accommodation:
> (a) Transport;
> (b) Accommodation;
> (c) Other tourist services not ancillary to transport or accommodation and accounting for a significant proportion of the package.

The article then goes on to describe the 'organizer' as "the person who, other than occasionally, organizes packages and sells or offers them for sale, whether directly or through a retailer".[38] Effectively, tour operators and travel agents who do business within the European Community will be bound by this definition and are therefore responsible to clients for most problems encountered with package holidays purchased there.

In other countries where this consumer protection legislation does not exist, claimants must simply rely on their rights at common law (contract, tort or agency). This of course raises issues of gathering evidence, deciding whom to sue and in which country, and can prove both costly and frustrating for litigants.

Mere Agent versus Principal Contractor

At common law, attempts to sue the tour operator/travel agent have fallen into two categories. In the first category (traditional), the tour operator is a mere agent who

---

[37] Ibid.
[38] For full text http://europa.eu.int/comm/consumers/policy/developments/pack_trav/pack_trav01_en.html

undertakes to arrange for the services to be performed by others. This defines the tour operator's services and responsibility narrowly.[39] In the second category (emerging), the tour operator is the principal contractor who undertakes to supply the services whether or not others perform them. This defines the tour operator's services and responsibility widely.[40]

In the preamble to *The International Convention on Travel Contracts*, there is recognition of the need to establish uniform provisions relating to travel contracts.[41] Hopefully, with time more states will become contracting parties to this type of Convention so that in future all travellers will be protected by a practical, systematic and consistent approach to the tour operator's role in the travel contract.

Other Issues

Image is everything in selling tourism products and services. The greatest deterrent for tourism is the perceived risk to health and safety while travelling in a country. In considering which countries to visit and which to avoid, tourists will weigh up risks like the threat of war, terrorist attacks, civil unrest, and weak rule of law, poor local building, safety and hygiene standards, harassment and theft by locals.[42] Only the few tourists who are seeking high adventure and are unconcerned about personal safety issues will ignore all the warnings.

The best antidote to this is for a country first to develop and enforce criminal and civil (consumer protection) laws, which encourage good behaviour from locals towards visitors and vice versa. Secondly, the tourist industry must set minimum standards of practice for its members so as to ensure that visitors have a positive experience, which meets their expectations. Thirdly, when all this is in place, the country should market itself as a safe haven with a friendly, secure and relaxed image.

Tourism versus Terrorism

Terrorism is now being regarded as the greatest threat to World peace. The international community has made attempts in the past to discourage terrorism,[43] but in more recent times it has redoubled its efforts. Two quite recent Conventions include the Convention for the Suppression of the Financing of Terrorism[44] and the Convention for the Suppression of Terrorist Bombings.[45] Just how committed the international community

---

[39] See cases like *Wall v. Silver Wings Surface Arrangements Pty. Ltd* (1981) (unreported) United Kingdom; *Craven v. Strand Holidays* (1981) 31 OR (2ED) 548 Canada; *Rockard v. Mexicoach* (1982) 689 FR 2d 1257 United States of America.

[40] See, for example, *Wong Mee Wan v. Kwan Kin Travel* (1995) 4 All ER 745; *P&O Steam Navigation Co and Ors v. Youell and Ors* (1997) 2 Lloyd's Rep 136.

[41] Brussels, 23 April 1970 http://www.unidroit.org/english/conventions/c-trav.htm

[42] Many countries publish travel advisories which warn their nationals of dangers of visiting risky countries.

[43] See for example, Convention Against the Taking of Hostages (12/79).

[44] Convention for the Suppression of the Financing of Terrorism (12/99).

[45] Convention for the Suppression of Terrorist Bombings (12/01/98).

really is about combating terrorism may be judged by the numbers of state parties who ultimately ratify these Conventions.

There is no doubt about the impact of war, terrorism and civil unrest on the tourism image of a country. For example, in Jordan, tourism has doubled since the country's 1994 peace treaty with Israel.[46] Egypt's tourism industry has received several serious blows in recent years. First the attack on tourists at Luxor in 1997 set the industry back severely and then it slumped again to about 30% of its previous capacity following the events of 11 September 2001. It is now slowly starting to recover. Similarly, the number of international visitors travelling to Lebanon has risen sharply since the end of its 15-year civil war but is still being held back by the neighbouring Israeli/Palestinian conflict.

Self Regulation

Apart from government intervention to ensure the safety and security of tourists at the national level, there are strategies that the tourist industry can adopt to help its image, support its products and services and reduce some of the inherent operational risks of the tourism business.

The most important of these strategies is the introduction of industry codes of conduct that set guidelines for best practice, minimum standards and quality assurance across a range of tourist products and services. In order to work, these codes require input and buy-in from all relevant industry players, and an awareness program to get the message out to the various target audiences including host communities and visitors.

Finally, in case all else fails, the tourist industry seeks to protect itself against large claims for personal injury, loss and damage through insurance. However, it has already been noted that insurers are reassessing many of the risks associated with tourist activities and attractions and charging proportionately larger premiums which some operators cannot afford.[47] Some adventure activities are becoming uninsurable.

Conclusion

This chapter has discussed a number of the current issues in travel and tourism law. New issues and challenges are emerging. All aspects of tourism are being affected, whether

---

[46] 'Peace is bringing tourism, and tourism is the propeller of our national economies,' said Jordanian Tourism Minister Akel Biltaji. But without security, that propeller could come to a stop.

[47] After 12 years, Col Skinner is selling his 27 bed backpackers' hostel in Katoomba, Australia. His Adventure Tourism business has cut white-water rafting from its program of rock climbing, abseiling and canyoning following premium rises of 300% this year. It could have been worse. He was quoted increases of 800%.

> 'I can't afford the hostel any more and if premiums keep escalating I will have to close the doors of my business, and the Adventure Tourism market, the great survivor of 11 September, will leave Australia too.' http://www.smh.com.au/news/0201/26/review/review5.html

it is passenger transport, traveller accommodation, activities and attractions, tour operators/travel agents or host communities themselves. There is a common interest in assuring the quality, health and safety of the services provided to tourists. There are few real winners when some part of the system fails and tourists suffer disappointment, personal loss or injury and seek compensation from those responsible or those legally liable. There is a need for greater leadership on these matters at the international level, particularly from the World Tourism Organisation, World Travel and Tourism Council, and the relevant international travel and tourism trade associations.

Chapter 8

Travel Agents' Health and Safety Advice

Jeff Wilks, Donna Pendergast and Leisa Holzheimer

Introduction

For many people, travel agents are seen as experts in all areas of travel information, including knowledge and advice about heath and safety issues. The expectation that travel agents will provide appropriate health and safety advice comes not only from their clients, but also from international health and medical organizations. For example, the World Health Organization (2000) proposes that:

> Travel free of health problems is in the interest of travel organizers and employees, as well as travellers. Travel agencies are encouraged to give their clients objective information on the hazards related to travel and their avoidance.

Similarly, a Consensus Statement on Providing Travel Health Advice issued by the International Society of Travel Medicine (1996) concludes ... "there is a need for greater emphasis on health advice for travellers *and that providing it should be part of the travel industry service*" (italics added).

In response to these expectations, travel agents point out that they are not doctors (Ivatts *et al.* 1999), nor do they have the time to become knowledgeable regarding medical advice (Schiff & Binder 1997). Travel agents are also conscious of the business implications of creating a negative impression of health and safety risks associated with certain destinations (Stears 1996) and their legal liability for incorrect information and advice (Ivatts *et al.* 1999). Even though many agents may still be reluctant to get involved in travel health and safety advice, the travel industry acknowledges that this is a topic that must be addressed (Foster 2001).

Provision of Health and Safety Advice

Surveys consistently show that travel agents are often consulted about health and safety issues. However, the proportion of clients seeking and/or receiving advice varies considerably, as do the standards and accuracy of the health and safety advice many travel agents are providing. In passing, a distinction should be made between health advice and safety advice, the latter being less likely to be discussed by travel agents (Ivatts *et al.* 1999) even though injuries account for more travel deaths and serious illness than infectious diseases (Hargarten & Gŭler Gŭrsu 1997).

Table 8.1 reveals that travel agents provide health and safety information inconsistently, and the quality and accuracy of that advice is variable. For some groups surveyed, travel agents are a common source of health and safety information (Grayson & McNeil 1988) while for others, travel agents provide little or no guidance (Lawton & Page 1997). In several contrived scenarios reported in the table, health and safety information was not provided unless solicited (Grayson & McNeil 1988; Grabowski & Behrens 1996; Harris & Welsby 2000; Lawlor *et al.* 2000). Of grave concern is that when given, the advice was generally inaccurate, with travellers advised to consume inappropriate medications and receive unnecessary immunizations.

Lawton & Page (1997) noted that while 16% of travel agents considered themselves to be the most appropriate source of health and safety information, 9% regard themselves as inappropriate for this role. In the study by Ivatts *et al.* (1999), more than half of the travel agents surveyed (56%) agreed that agents should provide travel health information, while a similar proportion indicated they would like to be more involved, but argued that 'not enough' travel health and safety information was available for agents.

Sources of Information

The argument that travel health and safety information is not readily available to travel agents cannot be sustained in today's business environment. In a recent review of sources of health advice given to travellers, Leggat (2000) noted that health advice is currently available from various groups within the travel industry, and from health professionals in travel clinics, hospitals, public health units, general practices or other centres. Travel health information can be accessed by telephone or through the Internet. Wilks (2003) highlights the importance and value of Australia's Department of Foreign Affairs and Trade (DFAT) travel information website http://www.dfat.gov.au/travel/index.html which monitors current activities in more than 80 overseas destinations. The advice and warnings provided on the site cover a range of current health and safety issues, including civil unrest, crime, infectious diseases, seismic activity and weather patterns. This site, and similar services provided by other governments, is free to access.

While Leggat (2000) concluded that travel agents and various travel industry publications would remain a central source of health information for travellers, a number of studies have raised concerns about reliance on brochures and publications. For example, surveys of travel brochures in the United Kingdom indicate that between 18 and 25% carried no health information (Cossar *et al.* 1993; Shickle *et al.* 1998).

Even if brochures and other forms of written information are widely available, current research shows that a large proportion of travellers will not seek out health and safety

Table 8.1: Research findings on health and safety advice provided by travel agents.

Year	Authors	Study Details	Findings
1988	Grayson & McNeil	Survey of 500 Australian travellers to Bali	87% of travellers received health advice from either their family doctor (75%), travel agent as sole source (12%) or both (59%). The advice given about immunizations and malarial prophylaxis frequently differed from recommendations set by the Australian government and the World Health Organization, with travellers being immunised unnecessarily and inappropriate medications recommended.
1996	Grabowski & Behrens	Covert survey of 202 travel agents in the United Kingdom	A scenario of travel from the United Kingdom to Kenya or India was staged. No health warnings of any kind were given by 61% of travel agents. After prompting, 71% gave general health advice, 63% mentioned malaria risk, and 60% gave advice to see a GP. Almost 10% did not mention health measures of any kind. The study reveals that travel agents provide health advice inconsistently and mention health risks only when prompted.
1997	Lawton & Page	Postal survey of 314 travel agents in New Zealand for outbound travel to Pacific Island destinations	60% of travel agents consider GP's to be the most appropriate source of health and safety information, while 16% believe travel agents fill this role. 9% regard travel agents as inappropriate sources. In terms of travel agents frequency of issuing health and safety advice: less than 20% always do so; 44% always or nearly always; 34% sometimes; and one fifth very rarely or never give advice. Around 66% always or nearly always advised travellers about vaccinations, with 63% always or nearly always recommending travellers ensure that food and water is clean. The study concluded that travellers are not well served by travel agents regarding health issues.

Table 8.1: Continued.

Year	Authors	Study Details	Findings
1999	Ivatts *et al.*	Survey of 145 travel agencies in Western Australia	56% of travel agencies give broad health and safety guidelines and recommend their clients consult a medical practitioner. While 56% agree that travel agencies should provide health information, the same proportion argue that it is not readily accessible. The most commonly used source is specialised travel medicine clinics (54%). 52% of travel agencies would like to be more involved in providing health information to their clients.
2000	Harris & Welsby	Covert survey of 80 travel agents and retrospective postal survey of travellers attending a travel clinic	A scenario of travel from the United Kingdom to Kenya was staged. Only 25% of travel agents gave correct advice for vaccinations and medications, with over half advising more vaccinations than required. 31% provided additional health advice or leaflets. Of the travellers attending a travel clinic, no less than 75% had been incorrectly advised by their travel agent about vaccinations required. The study concludes that travel agents have an implicit responsibility to issue travel advice and are failing to do so.
2000	Lawlor *et al.*	Covert survey of 5 travel agents in the United Kingdom for 3 scenarios — total of 15 incidents	Three scenarios of travel to Turkey, Kenya and Amsterdam were staged. Only one interaction of the 15 provided unprompted health and safety advice from the travel agent. When prompted, instead of referring clients to their GP's for immunisation advice, some provided inaccurate advice. Travel agents did not provide advice concerning food, drink, sun exposure and safe sex.

information for themselves (Peach & Bath 1998). It therefore becomes the responsibility of the travel agent to ensure that health and safety issues are addressed with the client. Various legal provisions now reinforce this obligation.

Legal Responsibility

The requirement having the greatest general impact on travel health and safety advice is that of the European Community Directive on Package Travel (1990). While details of the Directive are beyond the scope of this chapter, it is important to note that it prescribes a list of items that must be included in tour brochures and other written information (for details see Cordato 1999). Among these are information on health formalities for the journey and the stay. Vansweevelt (1999) points out that this means brochures must indicate, country by country, for which diseases it is recommended or compulsory to have a vaccination, and where travellers should take preventive medicine.

Under the Directive, the travel organiser and/or the retailer (e.g. travel agent) are liable to the consumer for the proper performance of their obligations. If a traveller suffers damage (e.g. a disease) due to insufficient or incorrect written information, the travel organiser and/or retailer will be liable. Since many carriers and accommodation providers in Australia and the Asia Pacific region are now being included in such obligations by virtue of their contractual arrangements with European Community package travel organisers, the Directive has wide ranging implications.

Cordato (1999) notes that in Australia, travel agents have particular duties under both common law and statute. As a lawyer, Cordato (1999: 218) recommends, "on any but the most mundane journeys or tours a travel medicine consultation should be recommended to the consumer". In addition, travel agents should highlight the medical, hospital and repatriation components in travel insurance policies sold to consumers.

Travel agents clearly have a legal 'duty of care' to their clients, and as such are expected to advise them fully and accurately about health risks and preventative measures, including suitable insurance policies (Foster 2001). In the area of insurance, a majority of travel agents appear to be comfortable proving information and advice (Ivatts *et al.* 1999).

Asia Pacific Destinations

Many destinations rely heavily on tourists being adequately prepared for their journey before leaving home. For most destinations in the Asia Pacific region this means insisting that visitors carry travel health insurance and have a pre-travel consultation with a medical practitioner (Wilks 2003). Travel agents are a logical group to provide this assistance, though as previously noted, recent studies show that health and safety advice provided by travel agents varies considerably (Table 8.1). For example, in their study of 314 New Zealand travel agents, Lawton & Page (1997) found that less than one third gave advice about malaria tablets for travellers to Vanuatu or the Solomon Islands.

Similarly, less than half the agents gave advice on the prevention of insect bites or care with local food and water, despite the very real risk of dengue fever from mosquito bites, and typhoid, hepatitis A and gastroenteritis being contracted through contaminated food or water in many destinations.

The prevention of unsafe sexual practices among travellers to some Asian destinations is also a priority. In one study, 60% of Australians visiting a travel medicine clinic prior to travel to Thailand reported an intention to have sex on their vacation (Mulhall *et al.* 1993). In another study, Rowbottom (1991) found that 50% of Australian men presenting to a sexually transmitted diseases clinic after returning from South East Asia had urethritis, 5% genital herpes, and 5% hepatitis B. While personal health issues such as sexual behaviour and sun protection, and personal safety issues such as road and water safety are very important in Asia Pacific destinations, these are topics that are least likely to be raised or discussed by travel agents (Peach & Bath 1998; Ivatts *et al.* 1999) or to be available in travel brochures (Peach & Barnett 2001).

Travel Warnings

As a result of the 11 September 2001 terrorist attacks in the United States, and the more recent bombing of tourist facilities in Bali, Indonesia, there is an increased expectation that travel agents will draw their customers' attention to government travel warnings issued against various destinations. In some cases, travel insurance companies have rewritten their policies since 11 September 2001 to specifically exclude coverage for terrorism and travel-supplier bankruptcy or default (Pacific Asia Travel Association 2002). Travelling to a destination that is currently the subject of a government travel warning may, in some cases, void the travellers' insurance policy. For this reason, awareness of government travel warnings is a critical factor in the health and safety advice provided by travel agents.

Table 8.2 lists the 24 countries that were subject to travel warnings from the U.S. State Department www.travel.state.gov/warnings_list.html as at 10 November 2002. A Travel Warning is issued when the State Department decides, based on all relevant information, to recommend that Americans avoid travel to a certain country.

The next level down are Public Announcements, which are a means to disseminate information about terrorist threats and other relatively short-term and/or trans-national conditions posing significant risks to the security of American travellers. These are made any time there is a perceived threat and usually have Americans as a particular target group. The countries and regions listed for Public Announcements as at 10 November 2002 are presented in Table 8.3. In the past, Public Announcements have been issued to deal with short-term coups, bomb threats to airlines, violence by terrorists and anniversary dates of specific terrorist events.

In addition to Travel Warnings and Public Announcements, the State Department issues Consular Information Sheets for every country of the world with information on such matters as the health conditions, crime, unusual currency or entry requirements, any areas of instability, and the location of the nearest U.S. embassy or consulate in the subject country.

Table 8.2: Current travel warnings issued by the U.S. State Department as at 10 November 2002.

Travel Warnings are issued when the State Department recommends that Americans avoid a certain country. The countries listed below are currently on that list.

Iraq — 10/31/02
Central African Republic — 10/31/02
Indonesia — 10/19/02
Cote d'Ivoire — 10/18/02
Libya — 10/7/02
Somalia — 8/23/02
Angola — 8/23/02
Pakistan — 8/12/02
Burundi — 8/9/02
Nigeria — 8/8/02
Israel, the West Bank and Gaza — 8/2/02
Sudan — 7/9/02
Afghanistan — 7/3/02
Colombia — 7/3/02
Congo-Kinshasa — 7/1/02
Pakistan — 6/26/02
Bosnia & Herzegovina — 6/4/02
Macedonia (Former Yugoslav Republic of) — 5/21/02
Liberia — 5/21/02
Lebanon — 4/29/02
Yemen — 3/18/02
Iran — 1/30/02
Algeria — 12/11/01
Tajikistan — 9/26/01

While in a slightly different format, Australia's Department of Foreign Affairs and Trade (DFAT) issues essentially the same information as the U.S. State Department, though it is interesting to note that at any one time some destinations may appear on one warning list but not another. For example, as at 10 November 2002 DFAT warned Australians not to travel to the Solomon Islands or the Southern Highlands and Enga Provinces of Papua New Guinea. These regional variations are important for travel agents since their legal obligations to their clients include having a high standard of product knowledge, which must be continually updated (Cordato 1999). Regular monitoring of government advisory services, including those of other governments, such as the Canadian Department of Foreign Affairs and International Trade http://voyage.dfait-maeci.gc.ca/destinations/menu_e.htm will ensure that travel agents are aware of current health and safety issues that may affect their clients.

Table 8.3: Current public announcements issued by the U.S. State Department as at 10 November 2002.

The State Department issues Public Announcements to disseminate information quickly about terrorist threats and other relatively short-term conditions that pose significant risks or disruptions to Americans. The current Public Announcements are listed below

Worldwide Caution: issued — 11/6/02, expires — 5/7/03
Middle East and North Africa Update: issued — 11/4/02, expires — 5/30/03
Philippines: issued — 11/3/02, expires — 1/10/03
South East Asia: issued — 11/2/02, expires — 5/1/03
Italy: issued — 10/31/02, expires — 11/15/02
Uzbekistan: issued — 10/31/02, expires — 04/28/03
East Timor: issued — 10/23/02, expires — 4/17/03
Malaysia: issued — 9/20/02, expires — 3/22/03
Turkmenistan: issued — 9/12/02, expires — 12/13/02
Nepal: issued — 9/3/02, expires — 12/19/02
India: issued — 7/22/02, expires — 11/20/02
Madagascar: issued — 7/15/02, expires — 11/15/02
Guatemala: issued — 7/3/02, expires — 12/1/02
Kyrgyz Republic: issued — 6/6/02, expires — 12/2/02

Continuing Education and Training

Reviews of the literature show that there is a clear gap between the expectations of clients, industry groups and government agencies as to the health and safety advice travel agents should provide, and agents' current willingness and ability to provide such advice. Some commentators consider that travel consultants "do not take their responsibility for travel health seriously" (Lawlor *et al.* 2000: 567), while others acknowledge that the travel industry has high staff turnover (Lea 1998) and therefore staff training in pre-travel health advice needs to be an ongoing consideration in all travel agencies.

As travel agents become increasingly concerned about possible legal action from injured, ill or disappointed clients, the importance of accurate and up-to-date pre-travel advice becomes more obvious. Grabowski & Behrens (1996) suggest that travel agents probably need training in providing health advice in such a way that it does not scare off potential travellers. Some agents still believe it is not their business to provide detailed health advice (Dawood 1989; Ivatts *et al.* 1999), so either they provide no information or dismiss the risks. For legal and professional reasons, this response is both inappropriate and dangerous.

While the health and medical professions have a major role to play in assisting travel agents with their continuing education, there is also an emerging role for universities, since comprehensive travel health and safety advice includes aspects of health, law, education and business. At a time when travel agents' booking commissions are being reduced, and an increasing trend toward fees for service is being observed in the industry (Rutledge *et al.* 2000), agents have an opportunity to make health and safety advice a positive contributor to their business. To do this effectively, several key practice elements need to be put in place:

- Travel agents should have at least one source of current health information available to them, and supplement this with regular checks on government travel advisories;
- Regular training must be available for staff so that they can identify high-risk travellers and high-risk destinations, as well as developing a reasonable knowledge of travel-related health and safety hazards; and
- Travel agents need to develop and follow set protocols when giving advice to clients.

Conclusion

Travel agents have a practical and legal responsibility to ensure travellers take suitable precautions when travelling overseas. In order to fulfil these responsibilities, travel agents need the support of their own governments, the wider travel industry and the destinations to which clients are travelling. Current literature shows that travel agents provide health and safety information inconsistently, and the quality and accuracy of that advice is variable. Continuing education and in-service training are needed to meet agents' legal responsibilities, and to add value in this area of customer service. Until recently, travel agents were most likely to receive requests about health advice from their clients. In the new millennium the challenge for agents will be for accurate and timely information about visitor safety.

References

Canadian Department of Foreign Affairs and International Trade. http://voyage.dfait-maeci.gc.ca/destinations/menu_e.htm

Cordato, A. J. (1999). *Australian travel and tourism law.* Sydney: Butterworths.

Cossar, J. H., McEachran, J., & Reid, D. (1993). Holiday companies improve their health advice. *British Medical Journal, 306*, 1070–1071.

Dawood, R. (1989). Tourists' health — Could the travel industry do more? *Tourism Management, 10* (4), 285–287.

Department of Foreign Affairs and Trade (DFAT). http://www.dfat.gov.au/travel/index.html

European Community Directive (1990). Council directive on package travel, package holidays and package tours. *Official Journal of the European Communities, L158*, 59–63.

Foster, J. H. (2001). Travel health and insurance: Fulfilling the duty of care. *Traveltalk Asia-Pacific*, (February–March), 26.

Grabowski, P., & Behrens, R. H. (1996). Provision of health information by British travel agents. *Tropical Medicine and International Health, 1* (5), 730–732.

Grayson, M. L., & McNeil, J. J. (1988). Preventive health advice for Australian travellers to Bali. *Medical Journal of Australia, 149*, 462–466.

Hargarten, S. W., & Güler Gűrsu, K. (1997). Travel-related injuries, epidemiology, and prevention. In: H. L. DuPont, & R. Steffen (Eds), *Textbook of travel medicine and healh* (pp. 258–261). Hamilton, Ontario: B. C. Decker.

Harris, C. B., & Welsby, P. D. (2000). Health advice and the traveller. *Scottish Medical Journal, 45*, 14–16.

International Society of Travel Medicine (1996). *North American charter for travel health. A consensus statement on providing travel health advice.* www.istm.org/consensus.html

Ivatts, S. L., Plant, A. J., & Condon, R. J. (1999). Travel health: Perceptions and practices of travel consultants. *Journal of Travel Medicine, 6*, 76–80.

Lawlor, D. A., Burke, J., & Bouskill, E. *et al.* (2000). Do British travel agents provide adequate health advice for travellers? *British Journal of General Practice, 50*, 567–568.

Lawton, G., & Page, S. (1997). Evaluating travel agents' provision of health advice to travellers. *Tourism Management, 18*, 89–104.

Lea, G. (1998). Doctors and the travel industry collaborate to improve health advice for travellers. *Communicable Disease and Public Health, 1*, 4.

Leggat, P. A. (2000). Sources of health advice given to travellers. *Journal of Travel Medicine, 7*, 85–88.

Mulhall, B. P., Hu, M., & Thompson, M. *et al.* (1993). Planned sexual behaviour of young Australian visitors to Thailand. *Medical Journal of Australia, 158*, 530–535.

Pacific Asia Travel Association (2002). September 11: The legal fallout. *Issues & Trends Pacific Asia, 7* (4), 1–3.

Peach, H. G., & Barnett, N. E. (2001). Health and safety information carried by travel brochures in Australia. *Medical Journal of Australia, 174*, 150.

Peach, H. G., & Bath, N. E. (1998). Australians travelling abroad without health and safety information: How many and who are they? *Asia Pacific Journal of Tourism Research, 3* (1), 64–74.

Rowbottom, J. (1991). Risks taken by Australian men having sex in South East Asia. *Venereology, 4*, 56–59.

Rutledge, J., Black, N., Clarke, A., & Bauld, S. (2000). *Travel agents in Australia: A review. CRC tourism work-in-progress report series: Report 5.* Southport, Queensland: Cooperative Research Centre for Sustainable Tourism.

Schiff, A. L., & Binder, M. (1997). Responsibilities and ideal interaction. In: H. L. DuPont, & R. Steffen (Eds), *Textbook of travel medicine and health* (pp. 10–13). Hamilton, Ontario: B. C. Decker.

Shickle, D., Nolan-Farrell, M. Z., & Evans, M. R. (1998). Travel brochures need to carry better health advice. *Communicable Disease and Public Health, 1*, 41–43.

Stears, D. (1996). Travel health promotion: advances and alliances. In: S. Clift, & S. J. Page (Eds), *Health and the international tourist* (pp. 215–234). London: Routledge.

U.S. State Department. www.travel.state.gov/warnings_list.html

Vansweevelt, T. (1999). The EC package travel directive and the health advice of the travel organizer, the travel agent and the physician. *Vaccine, 17*, S88–S89.

Wilks, J. (2003). Destination risk management in Oceania. In: C. Cooper, & C. M. Hall (Eds), *Regional tourism handbook: Oceania*. London: Channel View, in press.

World Health Organization (2000). *Note for travel organizers.* www.who.int/ith/english/note.htm

Chapter 9

Safety and Security for Destinations: WTO Case Studies

Jeff Wilks

Safety and Security at a National Level

According to a World Tourism Organization (WTO) survey of member countries on the Security and Protection of Travellers, Tourists and Tourism Facilities, national policy in tourism safety and security is usually the responsibility of the Ministry for Interior Affairs (World Tourism Organization 1994). The survey also revealed that there is a need to identify a clearer role and provide for more active involvement by national tourism authorities in safety and security issues.

While the WTO survey is now dated, recent reviews have confirmed that the effective management of tourist health and safety in the new millennium will rely considerably on the coordinated efforts of government departments at a national level for the benefit of destinations. Previous sections of this book have presented a number of perspectives on tourist health and safety, largely supported by research findings. This chapter offers another perspective through examination of selected case studies from a recent WTO review (World Tourism Organization 2003). The focus here is on national level safety and security initiatives for tourism in various countries.

According to the WTO, forming and coordinating partnerships are two fundamental requirements for organizing safety and security in tourism. These partnerships should involve:

- All national government agencies and departments, led by those concerned primarily with tourism;
- Tourism destination communities;
- Tourism industry representatives; and
- The media.

Managing Tourist Health and Safety in the New Millennium
Copyright © 2003 by Elsevier Science Ltd.
All rights of reproduction in any form reserved.
ISBN: 0-08-044000-2

Coordination and Partnerships — A National Tourism Council

One way to accomplish this coordination is to form a National Tourism Council. The Council should then organize a National Safety and Security Committee. In many countries, coordination is only carried out among government agencies. However, in tourism it makes sense to form a mixed-sector council with government and industry participants, since many of the actions can and should be implemented by the private sector. Government agencies and tourism industry sectors to consider for membership on the National Safety and Security Committee include:

- National tourism administration/tourist board;
- National police;
- Immigration;
- Judiciary;
- Customs;
- Transportation;
- Health;
- Foreign affairs;
- Civil defence;
- Airlines and transportation company associations;
- Hotel associations;
- Tour operators' associations;
- Travel agents' associations;
- Other travel and tourism representatives;
- Consumer groups;
- Retail trade organizations; and
- Tourism safety and security oriented research and documentation centres.

The WTO recommends that every country develop a national policy on tourism safety commensurate with the prevention of tourist risks (World Tourism Organization 1991).

National Tourism Safety and Security Plan

A National Tourism Safety and Security Plan is a logical consequence of the development of a national policy on this subject. Such a plan should address the following main areas:

- Identification of potential tourist risks according to types of travel, affected tourism sectors, and locations;
- Detection and prevention of offences against tourists;
- Protection of tourists and residents from illicit drug trafficking;
- Protection of tourist sites and facilities against unlawful interference;
- Establishment of guidelines for operators of tourist facilities in the event of unlawful interference;
- Responsibilities for dealing with the press and other media, at home and abroad;

- Information to be provided to the international travel trade on safety and security issues;
- Organization of crisis management in the event of a natural disaster or other emergency;
- Adoption of safety standards and practices in tourist facilities and sites with reference to fire protection, theft, sanitary and health requirements;
- Development of liability rules in tourist establishments;
- Safety and security aspects of licensing for accommodation establishments, restaurants, taxi companies, and tour guides;
- Provision of appropriate documentation and information on tourist safety to the public, for both outgoing and incoming travellers;
- Development of national policies with regard to tourist health, including reporting systems on health problems of tourists;
- Development of tourist insurance and travel assistance insurance; and
- Promotion, collection and dissemination of reliable research statistics on crimes against travellers.

Implementation of a safety and security plan will be enhanced by setting up a database of model programs, useful practices and reliable data on problems affecting tourists. Accurate, neutral and reliable data is important for improving the response of tourism authorities to safety and security problems.

In the recent WTO review of tourist safety and security issues (World Tourism Organization 2003) no one destination was found to have in place all the proposed measures and systems of a National Tourism Safety and Security Program listed above. However, some countries (such as South Africa) had developed key elements of a national program and the potential for expanding their partnerships in other areas. The review identified a number of outstanding initiatives in different countries. Some are presented here as case studies. The point to be highlighted is that a truly national program in tourist safety and security takes time and effort to build.

Case Study: Tourist Safety and Security — The South African Experience

In South Africa the government approaches the question of tourist safety and security from the point of view of a partnership with various agencies, led by the Tourism Safety Task Group (TSTG). The Task Group consists of the Department of Environmental Affairs and Tourism, the South African Police Service (SAPS), the South African Tourism Board (Satour), the Tourism Business Council of South Africa (TBCSA), Business Against Crime (BAC), the Department of Foreign Affairs, and the nine provincial tourism departments (Mokaba 1997). More recently, safety and security have been included in a national tourism initiative.

The SA Welcome Campaign

On 8 December 1999, the SA Welcome Campaign was launched, sensitising South Africans to the needs of the international traveller. It also encourages local tourism, national pride and caring for the environment. The campaign began with advertising in newspapers and on outdoor billboards, radio, cinema and TV to introduce the public to a flag-draped 'Welcome' character, with arms outstretched in greeting (Figure 9.1). With the 'Welcome' the core message is that 'Tourism is everyone's business.' www.sawelcome.com/about/index.html

Legal, Regulatory and Judicial Frameworks

It is primarily a national responsibility to provide the legal, regulatory and judicial frameworks that underpin tourism safety and security. A useful approach employed by a number of countries to evaluate their legal situation regarding tourism safety and security is to organize a systematic review of tourism safety and security problems and contrast them with the legal remedies and procedures available to deal with them.

Tourism officials can also make certain that judges, law enforcement officials, legislators, lawyers, and other government officials, as well as the tourism operational sectors are invited to participate in this systematic review. The objectives are to identify gaps in the laws or regulations and possible corrective measures for dealing with tourism safety and security issues. Results of such a review might suggest a more flexible approach by the judiciary is required, or safety standards or liability rules need to be articulated more carefully, or the country needs to create conditions for basic, affordable insurance schemes to cover travel and tourism product-related risks.

An excellent example of change being made to a legal framework to address a tourist issue is that of Thailand's laws on prostitution.

Case Study: Thailand Legislation

In April 1996, the Royal Thai Government passed stringent anti-prostitution laws with the most severe penalties reserved for those involved in child prostitution. Now customers, procurers, brothel owners, those who force children into prostitution and even parents, face long prison sentences as well as large fines. The penalties for a customer (including a tourist) who buys sexual service from a child prostitute under Thailand's Prostitution Prevention and Suppression Act B.E. 2539 (1996) are:

- Two to six years jail if prostitutes are under 15 years of age, and a fine of 40,000–120,000 Baht; and
- One to three years jail if prostitutes are between 15 and 18 years of age, and a fine of 20,000–60,000 Baht.

The Tourism Authority of Thailand (TAT) fully supports these tough new measures and asks visitors to their website www.tat.or.th/visitor/child_pros.htm to help end child

Welcome

For every eight tourists who visit our country, one job is created.

These are some of the things you can do
to help tourists feel more welcome:

- Give a friendly smile
- Stop and talk
- Be positive
- Be proud of our facilities
- Keep our environment litter free
- Talk about our fantastic food and wine
- Be proud of our different cultures and customs
- Say no to crime
- Be proud of being South African

Remember, tourism is everybody's business.

Figure 9.1: SA welcome campaign.

prostitution. TAT requests anyone who knows of any organization, operator or individual who offers sex tours to Thailand, particularly those involving children, to please report them to the local TAT office or contact TAT by e-mail.

Stopping Organized Crime

At the national level, tourism authorities can support the actions of police in combating organized crime and terrorism by firstly supporting efforts to make it more difficult for criminals to enter a country or to carry out unlawful acts. This means supporting controls at airports and other ports of entry that allow legitimate travellers to enter the country with minimum obstacles, while effectively keeping out criminals and their weapons. It involves increasing alertness at hotels and other accommodation establishments, as well as at car rental agencies and other means of transportation.

A second element in the approach is to identify the precise nature of criminal acts and design appropriate information campaigns for tourists and the domestic and international travel trade. Organized violent crime, in particular, is usually highly targeted, even though it may appear random. Specific counter-measures or instructions to the travel trade and to tourists can result from an analysis of specific problem situations. For example, in response to ongoing problems with crime related to car and coach travel in Mexico, the government instituted a highway service called 'The Green Angels' and supported this initiative with travel safety information materials for tourists.

Case Study: The Green Angels — Mexico

Visitors who find themselves in trouble while driving in Mexico can call the Ministry of Tourism's hotline on (91)(5) 250-8221/8555 extension 130/297 to obtain help from the 'Green Angels,' a fleet of radio dispatched trucks with bilingual crews that operate daily. Services include protection, medical first aid, mechanical assistance and basic supplies. Visitors are not charged for services, only for mechanical parts, fuel and oil. The Green Angels patrol daily, from dawn until sunset. Visitors who are unable to call them are advised to pull off the road, lift the hood of their car and wait for the roving patrols. http://travel.state.gov/tips_mexico.html

Policing in Tourism

Sixty-seven countries replied to the World Tourism Organization (1994) survey on Security and Protection of Travellers, Tourists and Tourism Facilities. Several of the survey questions concerned police services for tourists. The survey findings were reported at an Experts Meeting in Madrid (World Tourism Organization 1994) where delegates recommended, inter alia "the allocation of adequate resources to the courts, the police, and the public and private security forces for the protection and general well-being of travellers and tourists".

Based on the 1994 survey results, a select group of countries was approached in 1995 to obtain more detailed information on police work and tourism. The responses of 24 countries confirmed the importance of policing for tourism (Handszuh 1997). In some countries the regular police force deals with tourist matters, whereas in others specialist tourist police services have been formed. For example, in Greece the Tourism Police are an integral part of the Hellenic Police (ÅËÁÓ), consisting of men and women especially trained and competent to offer information and assistance to tourists. Officers all speak foreign languages and can be recognized by the shoulder flash 'Tourism Police' on their uniforms. Tourism Police operate a 24-hour emergency telephone line (dial 171 from anywhere in Greece) http://www.gnto.gr/1/01/0112/ea0112000.html A similar specialist police unit operates in Malaysia.

Case Study: Malaysia—The Tourist Police Unit

The Royal Malaysian Police established the Tourist Police Unit on 22 February 1988. There are approximately 400 selected officers, men and women, who have a good command of English and are given further training in public relations and communication skills; foreign languages (Japanese, Thai, Arabic); first aid; courtesy, tourism and policing (investigation, traffic control, leadership and command, and management and prosecution). The Tourist Police uniform is identified by its chequered lining on the cap/helmet and the logo (i) on the right side pocket.

Prevention of Drug Trafficking

The prevention of drug trafficking is another area where national tourism authorities may be called upon to work closely with police services and international agencies such as the World Customs Organization, Interpol, and the United Nations Office for Drug Control and Crime Prevention (World Tourism Organization 2003). Trafficking in illicit drugs violates national and international drug abuse laws. Such practices jeopardize the health, safety and security of tourists and travellers and of host country residents alike, and give rise to other petty and organized crime activities.

Tourists often want to try new things while on vacation. Illicit drugs may be one of them. Tourists must be warned of the legal penalties associated with illicit drugs, whether as smugglers, buyers, sellers, or users. For example, many tourists are unaware that the drug laws of the host country are applicable, not those of the tourist's citizenship.

For most countries the U.S. State Department's Consular Information Sheets are a very good source of information. They are available on-line. For example, the Malaysian site http://travel.state.gov/malaysia.html reminds travellers that while in a foreign country they are subject to that country's laws and regulations, which sometimes differ significantly from those in the tourist's home country. Persons violating the law, even unknowingly, may be expelled, arrested or imprisoned. The Malaysian Criminal Code includes a provision for a sentence of caning for certain white-collar crimes, including criminal misappropriation, criminal breach of trust and cheating.

Malaysia strictly enforces its drug laws. Malaysian legislation provides for a mandatory death penalty for convicted drug traffickers. Individuals arrested in possession of 15 grams (1/2 ounce) of heroin or 200 grams (seven ounces) of marijuana are presumed by law to be trafficking in drugs.

Assistance for Victims

While it is not always possible to prevent crime from occurring, WTO's *Recommended Measures for Tourism Safety* (World Tourism Organization 1991) proposes that countries cooperate in ensuring that a tourist who is the victim of an unlawful act receives all the necessary assistance possible.

An excellent example of a national program offering assistance to travellers is the Tourist Victim Support Service in Ireland. Part of the national Victim Support Organization, with 40 branches and over 500 volunteers, the tourist program works closely with the Irish police (the Garda).

Case Study: Tourist Victim Support Service, Ireland

Tourist Victim Support Service (TVSS) is a unique service for tourists who have been robbed, attacked or otherwise victimized during their stay in Ireland.

Tourist Victim Support will:

• Provide emotional and practical support;
• Arrange accommodation/meals if needed;
• Liase with embassies;
• Help to replace stolen travel tickets;
• Cancel credit cards and travellers cheques;
• Act as a base for money transfers;
• Arrange discount for windscreen repairs;
• Arrange transport for stranded tourists;
• Assist with language difficulties;
• Address medical needs;
• Offer telephone and fax facilities; and
• Offer a 24-hour free phone number 1800 661 771.

Tourist Victim Support Service (TVSS) does not:

• Offer financial assistance;
• Replace lost items; or
• Offer insurance or legal advice.

Referrals to TVSS come through the Garda. The severity of the crime does not necessarily determine the need for a referral. The main types of crimes dealt with are pick pocketing, thefts from cars and handbag snatches.

TVSS is widely acclaimed and enjoys the support of the Minister for Tourism, Sport and Recreation, the Minister for Justice, Equality and Law Reform, the Garda Commissioner, Bord Failte and many tourism related industries. Tourist Victim Support Service is based in Garda Headquarters, Harcourt Square, Dublin courtesy of the Department of Justice and the Garda Siochana. The success of the service depends on extensive support coming from a cross section of the tourism related industry by way of direct funding and benefit in kind.

Monitoring of Tourist Safety and Security Issues

Large-scale industry surveys are a very useful tool for monitoring tourist safety and security issues, since they focus attention on specific areas where problems occur. At the WTO seminar on Tourist Safety and Security held in Warsaw in 1997, results of a major survey (380,000 interviews) by the European Travel Monitor were presented (Anonymous 1997). An important finding was that more than eight million Europeans on trips abroad (or 3% of all European travellers) were victims of serious crime. Figure 9.2 shows that the most common criminal offences were car break-in/theft, handbag theft, fraud on money exchange, and theft of cash, credit cards and cheques.

The implications of these findings for tourist destinations are that general safety information should be made readily available to travellers, especially information on currency, exchange rates and average prices of common purchases; and the police need to be a central partner in tourist safety and security committees.

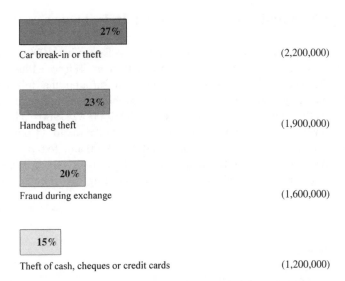

| 27% | | |
| Car break-in or theft | | (2,200,000) |

| 23% | | |
| Handbag theft | | (1,900,000) |

| 20% | | |
| Fraud during exchange | | (1,600,000) |

| 15% | | |
| Theft of cash, cheques or credit cards | | (1,200,000) |

Figure 9.2: Crimes against European tourists in 1996.

Source: European Travel Monitor, IPK International.

Based on the European Travel Monitor survey findings, recommended information for tourists should include:

- Exchange rates;
- Design of banknotes;
- Common rules of behaviour;
- Places to avoid travelling;
- Safe places to leave luggage;
- Average prices of common purchases;
- Emergency telephone numbers;
- The need to report crimes before tourist returns home; and
- The importance of keeping copies of documents.

General Support for Tourists

Safety and security are only two elements, though very important ones, in complete quality tourism service at any destination. As part of a general communications program, tourists should be able to easily receive assistance on a variety of matters, including safety and security. This is particularly important if the destination is one where language barriers may present additional problems. An excellent program in the area of general tourism support is the Italian National Call Centre, which is recommended as a model for all tourist destinations.

Case Study: National Call Centre 'More Italy Than Ever'

The National Call Centre has been operational since December 1999. It was created following an agreement between the Tourism Department, Telecom Plus International SpA and ENIT. The Call Centre can be accessed through the toll free number 800-11-77-00 if calling from any public, private or mobile phone in Italy. If calling from abroad, access is through a local number in Rome: 0039-06-87419007. The Call Centre is operational every day, including holidays, from 8:00 a.m. to 11 p.m., except for Christmas day and Easter when it can be accessed from 8:00 a.m. to 6 p.m. Multilingual operators are available to give information (of certified quality since it is issued from public institutions and territorial bodies) in Italian, English, French, German and Spanish on the following topics:

- Security;
- Health care;
- Events and shows;
- Museums and exhibitions;
- Public transportation and traffic;
- Useful numbers; information point; tourist and religious assistance facilities;
- Accommodation (fax on demand).

Tourist Safety and Security and Destination Image

Even taking into account the terrorist attacks of 11 September 2001 in the United States, organized crime and terrorism are still at the bottom of the list of real threats to tourists, but they are near the top of the list for travellers' perceptions about safe destinations and where they will go on vacation. A person at home understands their day-to-day risks and lives with them. But that same person as a tourist does not want to be worried about uncontrollable risks from crime and violence when on vacation, all the more so if the language is strange, the surroundings unfamiliar and official procedures are hard to understand.

Not dealing with potential customer perceptions of safety, whether the perceptions are accurate or not, can result in lost income and opportunities for a destination. For example, to gauge reaction to the 11 September 2001 bombings and how it might affect travel plans, the United Kingdom travel magazine Wanderlust surveyed their readers. Results showed that these committed travellers were unlikely to cancel their planned trips, but many were changing destinations. Table 9.1 shows the places where respondents were more or less likely to now travel.

While Australia and New Zealand were perceived as safe destinations in the Wanderlust survey, the recent murder of English backpacker Caroline Stuttle in Bundaberg (Herde 2002) and the extensive international media surrounding her death are a timely reminder that even safe destinations like Australia must constantly nurture and protect their reputation.

Ironically, a review of all deaths of overseas visitors in Australia between 1997 and 2000 reveals that 76% were from natural causes and, apart from fatalities in the Childers Backpacker fire in 2000 (Queensland Fire and Rescue Authority 2000), there were only three homicides involving visitors in the country during that four year period (Wilks *et al.* 2002). The timely availability of these figures would have greatly assisted tourism authorities defend the country's reputation for safety. However, Australia, like many other countries, is still reluctant to fully commit to a national strategy on tourist safety and security.

Conclusions

In summary, this chapter presented an approach to tourist safety and security at destinations that highlights the lead role of national government agencies. The approach is endorsed and promoted by the World Tourism Organization. A global review showed that no one destination has in place all the WTO proposed measures and systems of a National Tourism Safety and Security Program. However, some countries had some very good initiatives, and these were presented as case studies. In managing tourist safety and security issues in the new millennium, partnerships between tourism authorities and police services will be critical. Even in countries that have a good safety and security record there is no room for complacency. One or two negative incidents with international media exposure can substantially damage the reputation of a destination

Table 9.1: Wanderlust survey results.

THE LOSERS

Region	Specific Country
1. Middle East −68.5%	1. Pakistan −34.4%
2. North Africa −17.9%	2. Indonesia −20%
3. India − 8.4%	

THE WINNERS

Region	Specific Country
1. Europe + 28.9%	1. Australia +7.0%
2. Latin America + 15.8%	2. New Zealand +5.4%
3. South Africa +4.2%	

(Wanderlust www.wanderlust.co.uk)

and impact on visitor numbers. Management therefore must be based on active partnerships between government agencies, the tourism sector and the media.

References

Anonymous (1997). WTO safety task force. *IACVB News*, (November), 26.

Handszuh, H. (1997). *Policing in tourism for visitor and resident protection*. Report from a WTO survey. Madrid: World Tourism Organization.

Herde, C. (2002). Qld: British backpacker robbed and thrown to death off bridge. *AAP General News*, (11 April).

Mokaba, P. R. (1997). Speech delivered by South African Deputy Minister Hon. Peter R. Mokaba, MP at the WTO Seminar on Tourism Safety and Security, Addis Ababa, Ethiopia, 25 April.

Queensland Fire and Rescue Authority (2000). *Building fire safety in Queensland budget accommodation*. A report resulting from the fire in the Palace Backpackers Hostel, Childers. Brisbane: Queensland Fire and Rescue Authority. Available at www.fire.qld.gov.au/about/publications.asp

Wilks, J., Pendergast, D. L., & Wood, M. T. (2002). Overseas visitor deaths in Australia: 1997–2000. *Current Issues in Tourism*, 5, 550–557.

World Tourism Organization (1991). *Recommended measures for tourism safety*. Madrid: World Tourism Organization.

World Tourism Organization (1994). *Collected papers from the WTO experts meeting on tourist safety and security, 11–12 April 1994*. Madrid: World Tourism Organization.

World Tourism Organization (2003). *Safety and security in tourism: Partnerships and practical guidelines for destinations*. Madrid: World Tourism Organization, in press.

Section 4

Selected issues

Chapter 10

International Tourists and Transport Safety

Bruce Prideaux

Introduction

Travel in the twenty first century is more comfortable, reliable, cheap, accessible and importantly, safer than at any other time. Earlier, travel was dangerous as boats sank, highway travellers were robbed or even killed and the risk of disease or accidental injury was an ever-present travel companion. Journeying from a point of origin to a destination was to endure a state of travail. While travel is now safer, it is not without risk. Table 10.1 provides some examples of these risks. Accidents do happen and travellers suffer injuries and death, although on a minuscule scale compared to the total number of journeys undertaken. Yet, in this era, where instant communications can beam graphic images of carnage into homes, the risk of travel, while small, is highly visible and accidents potentially influence tourists' travel decisions. Equally, there is the impact on tourists when injury results in lost travel time and itinerary changes (May 1989).

Compared to deaths caused by disease, cardiovascular incidents and other health conditions, transport accidents often involve relatively large numbers of people who suffer injury or death in single incidents. Surprisingly, the available literature on international tourists and transport safety is limited. For example, in the area of road transport, Watson et al. (1999) noted that while road safety is an important issue for tourism authorities, it, along with many other areas of tourist health and safety, has not been given a sufficiently high priority, and mishaps are largely left to be handled by insurers.

The tragic events that unfolded in the United States on 11 September 2001 trumpeted the impact of major transport disasters on tourism. While these were the actions of a terrorist organisation, the images of large passenger aircraft flying into Towers One and Two of the World Trade Center graphically illustrated the vulnerability of tourists to all forms of transport disasters, particularly where aircraft are involved. Continuing incidents involving public transport may reinforce the disquiet held by many travellers. While public transport accidents involving large numbers of people are usually widely reported, these are often not the leading cause of transport injury deaths of international tourists. Privately driven motor vehicle accidents are responsible for more injury deaths

Managing Tourist Health and Safety in the New Millennium
Copyright © 2003 by Elsevier Science Ltd.
All rights of reproduction in any form reserved.
ISBN: 0-08-044000-2

Table 10.1: Public transport accidents between 15 April and 26 May 2002.

Date of Incident	Mode of Transport	Location	Injuries	Incident Cause
15 April	Air China B 767	Busan South Korea	129 killed	Unknown
3 May	Coastal ferry	Chandpur Bangladesh	328 killed possibly more	Sunk during storm, possible overloading
4 May	Coach	Pyongyang North Korea	9 killed 11 injured	Accidental
4 May	EAS Airlines BAC-111	Kano Nigeria	100 plus killed	Crashed after take-off
7 May	China Northern Airlines MD-82	Dalian China	112 killed	Reported fire in the aircraft's tail
7 May	Egypt Air Boeing 737	Tunis Tunisia	23 killed 33 survived	Unknown incident during landing approach
8 May	Bus	Karachi Pakistan	27 killed	Suicide bomber (probable terrorist link)
10 May	Train	Potters Bar United Kingdom	7 killed 100 injured	Derailment- under investigation
11 May	Train	Lucknow India	10 killed 100 injured	Derailment- under investigation
25 May	China Airlines B 747	Taiwan Straights	225 killed	Apparent mid-air explosion
26 May	Train	Mozambique	205 killed 400 injured	Rail accident — derailment

than public transport (Wilks *et al.* 1999), but because they usually only involve small numbers of deaths in any one accident, these attract little or no media attention.

This chapter examines transport safety issues that affect international tourists from the perspectives of travel via public transport and travel by means other than public transport. To highlight factors leading to accidents that involve international tourists travelling by private transport, the incidence of motor vehicle injuries and fatalities in Australia are examined. The measures that have been adopted by the authorities to reduce accidents are explored and an example of a tourist safety program recently introduced in Australia is provided.

In recent decades, the opportunity for international travel has increased as a consequence of improved transport technology and improved access to most parts of the world. Transport safety has become a significant issue and accidents involving public transport usually attract considerable media attention. Many governments have responded to concerns about public transport safety by establishing a network of agencies to oversee safety regulations. For the tourism industry, the issues of safety and harm minimisation have also become increasingly important. According to Cartwright (2000: 160) "the risk of harm is the probability that it will occur as a result of exposure to a hazard" where a hazard is described as ". . . a set of circumstances that could lead to harm". Travel by public transport as well as by self-drive cars and even walking exposes travellers to a range of hazards, many of which they have little control over. However, visitors arriving in a destination do not expect to be injured and normally anticipate a holiday experience defined by relaxation and good health. Hunter-Jones (2000) observed that visitors often exhibit a high degree of apathy to risk, leading to a failure to take responsibly for personal safety. Visitors who may not be familiar with local laws and conditions may unknowingly expose themselves to situations that are dangerous and may result in serious injury or death. For a destination where a pattern of visitor injuries has emerged, failure to improve visitor safety may lead visitors to divert to competing destinations.

Tourist Travel Patterns

From a tourism perspective, transport is defined as the operation of, and interaction between, transports modes, ways and terminals that support the tourism industry through the operation of passenger and freight flows into and out of tourism destinations. The transport system links tourism generating regions (TGR) to tourism destination regions (TDR) and provides transport services within destinations. The transport system also exerts considerable influence over the volume of tourist flows between TGRs and TDRs, the developmental stage of the destination's tourism industry (Prideaux 2000) and the rate at which the destination is able to grow. For example, prior to the introduction of jet passenger services, travel to the Caribbean was difficult and slow and the region was regarded by Europe's tourism industry as being a destination located in a distant periphery. During the 1960s, the introduction of scheduled jet passenger services between Europe and the Caribbean also altered the region's status from one of a far to a near periphery. The rapid growth of air services, particularly after the introduction of commercial jet passenger services from 1960 onwards, has had a similar impact on many TDRs. Some remote TDRs remain (e.g. St Helena in the Atlantic Ocean and the Antarctic) and can only be reached after a comparatively lengthy sea voyage.

Figure 10.1 illustrates the tourist transport system based on the journey profile of various groups of tourists. Between the tourist generating region and the tourist destination region lies a transit zone, which is described as the region and places through which departing and returning travellers pass during their travel. In this transport zone, smaller tourism areas relying on the advantage of intervening opportunity (the creation of alternative lower cost tourism regions situated between major TGRs and TDRs) have emerged to service tourists enroute in the transit route. Once in the TDR tourists often

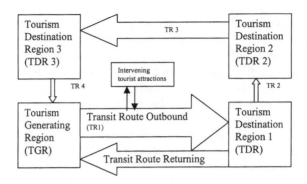

Figure 10.1: The opportunities for multiple transit routes to connect tourism
destination regions with a tourism generating region.

undertake further localised journeys and may also travel to other TDRs in a form of circuit tourism.

Globally, the largest TGRs are located in North America, Europe and more recently in North Asia (Japan, Korea, Taiwan and the People's Republic of China). Nationally, metropolitan cities operate as TGRs for other regions in the country. The introduction of wide bodied passenger jets, very fast rail services and high speed road systems has enabled TDRs to draw tourists from increasingly distant TGRs, leading to a lengthening of the transit route thereby increasing the scope for communities along the transit zone to develop their tourism sector. Once in a TDR, tourists have a variety of transport options available to them including public transport, self drive, cycling and walking.

Transport Safety in Tourism Destinations Regions

The impact of safety issues is reported to exert a strong influence on consumer's destination choice (Sonmez & Graefe 1998) and policy-making implications. Prideaux & Master (2001: 24) described destinations as "complex structures that exist in several senses: a spatial sense that has a physical presence, an economic sense that entails trade and profit, an ecological sense that draws together the unseen and seen world of nature, a social sense where people's perceptions and feelings colour their sense of satisfaction and enjoyment, and a historic sense that weaves the past into the present with the promise of the future. Together, the collective multiplicity of factors provides a unique tourism experience". For a destination to succeed long term, it must incorporate into its structure the ability to consistently deliver a safe and harm free holiday experience. Just as a destination's uniqueness draws the attention of tourists towards it, the inability to provide a safe experience will drive visitors away.

Unfortunately, not all destinations have the capacity to provide international travellers with a safe travelling environment. Such an environment is best described as the expectation of low travel risk, largely free of a sense of fear of becoming an accident statistic or suffering from criminal activity. In many developing nations road safety is not accorded a high priority and international tourists may face a higher degree of risk from injury while travelling by either public or private transport than in their home countries. For example, in Indonesia driving can be dangerous and one widely respected guidebook published by Lonely Planet (Turner *et al.* 1997: 159) gave readers the following advice "Driving yourself is not much fun in many parts of Indonesia. It requires enormous amounts of concentration and the legal implications of accidents can be a nightmare — that is if you survive an angry mob should anyone be hurt". On driving in Bali, the same publication (Turner *et al.* 1997: 386) stated, "Driving is risky. If you have a serious accident, it may be wise to go straight to the nearest police station rather than confront an angry local mob. If it's a minor accident, you may be better off negotiating a cash settlement on the spot than spending the rest of your holiday hassling with police and lawyers". Similar advice is given in many guidebooks for parts of Asia, Africa and Latin America. Although it is apparent that safety issues raised by guidebooks and media reports influence tourist behaviour in destinations, Mawby (2000) noted that relatively little is known about how incidents of this nature and other threats to safety affect the choice of holiday destinations.

An additional problem faced by destinations and individual enterprises in a very competitive international market place is the need to offer tourism products at the lowest possible price. Hunter-Jones (2000) observed that the United Kingdom market is primarily interested in 'sun, sea and sand' package tour products at the cheapest possible price, leaving little scope for the inclusion of safety measures that may add to the cost of the package with little additional advantage perceived by the purchaser. Commenting on the problem of the cost of safety verses destination competitiveness, Prideaux & Master (2001: 25) observed "that there are trade-offs between the cost of tours and visitor safety, with the emphasis on cost [being] a matter of some concern meriting further attention".

Transport Safety

Travel risk is broadly defined as the possibility of mishaps occurring during the process of travel to, from, or in a destination (Tsaur *et al.* 1997), and includes a range of factors such as disease, crime and transport related injury. During travel the tourist may use a variety of transport modes, usually defined as air, sea, rail and recently space. Some trips will be by public transport, others will be independent and may include self-drive components, limited water travel, cycling and walking. Although not usually thought of as a form of transport, walking tours and hiking are an important component of tourism. If a public transport option is selected for a journey, the only control that a tourist may exercise is the right of refusal to travel if the means of transport appears to pose some level of risk. This type of refusal to travel by public transport was exercised following the 11 September 2000 terrorist attack. The United States Government was forced to respond with a range of measures to enhance the safety of air travel through additional safety policies. Other incidents of death or injury involving travellers using public transport have rarely invoked such a dramatic government response as occurred in the

United States during the latter part of 2001. Actions to prevent further terrorist attacks included closing all United States civil airports for three days, deployment of armed soldiers to guard airports and when flying resumed, rigorous inspections were conducted of all passengers including random body searches.

When tourists elect to self-drive, cycle or walk they assume considerable control over their own safety. Unfortunately, unfamiliarity with new environments raises a number of safety concerns that may be ignored by the tourist and may be a major contributing factor to motor vehicle, cycling and walking injury deaths. Research (Ellis 1999) discussed later highlights the dangers that tourists may encounter with local traffic hazards when visiting Australia.

Governments respond to transport accidents according to their scale and severity. Where public transport is involved, the authorities in most jurisdictions will initiate some form of official investigation aimed at identifying the cause of the accident, and in most cases make appropriate changes to reduce the likelihood of further accidents. However, where the accident involves individuals or small groups of travellers, such as cyclists, walkers or motorists, the response may be on a lower scale and opportunities for intervention to reduce future accidents may be missed.

Public Passenger Transport

The ability of the transport system to move large numbers of passengers over long distances in relative safety and at an affordable price has been an important element in the development of mass tourism. In the air the introduction of commercial passenger jets such as the Boeing 707 made transoceanic travel possible for large numbers of people. It was responsible for the development of new destinations located far away from the main generating regions of North America, Europe and in recent decades North Asia. As the number and size of aircraft increased the potential for accidents also increased. Globally, the number of aviation accidents is relatively low although the number of causalities that occur during a single accident may be large. Similar observations about increased passengers carried and reduced accident rates can also be made about railways, shipping and coaches.

Accidents involving public transport often gain wide coverage in the global media and indicate a fascination by the media, particularly where foreigners are among the fatalities or injuries. To illustrate this observation, Table 10.1 lists the major public transport accidents and attacks involving public transport 15 April–26 May 2002 as reported in the Australian print media.

Aviation provides an example of safety management with the public sector in conjunction with operators. According to Hanlon (1996), market forces alone cannot be expected to enforce airlines to implement a consistent degree of attention to safety standards, despite an airline's strong commercial reasons for safe operations. In a rational world, it is unlikely that people will want to fly with an unsafe airline, particularly where choice exists. Surprisingly, airlines at times will ignore this reality and consistently operate aircraft in an un-airworthy condition or adopt unsafe operational practices. In 2001 Ansett Airlines, a domestic carrier in Australia, had its fleet of Boeing 767 aircraft grounded because of the airline's failure to follow aircraft manufacturers' recommendations on maintenance and inspection schedules. While the airline's maintenance shortfalls did not result in the loss of an aircraft, these problems were symptomatic of deeper financial problems that led to the airline's demise in September 2001. Maintenance budgets are frequently one of the areas that airlines experiencing financial problems look at to minimize costs.

However, not all airlines have improved their safety records in recent years and where a company continues to suffer a poor safety record it can expect to suffer from increased costs through rising insurance premiums, declining profits and reduced demand. China Airlines, the major international carrier servicing Taiwan, has a poor safety record with 12 fatal crashes and many serious non-fatal incidents between 1969 and 2002 (Chung-yan 2002). The airline's poor safety record has affected profit performance. After experiencing a NT$2.1 billion loss in 1998 the airline introduced a number of strategies to improve its safety record and after two accident-free years saw profits improve to NT$2.9 million in 2000 (Tak-ho 2002). The loss of a Boeing 747 on 25 May 2002 with 225 deaths is likely to have a severe impact on the airline and according to Tak-ho (2002) will result in decreased sales revenue, a cut back in ambitious fleet expansion plans, huge compensations claims and falling profits.

Recognition of the need to enhance standards of aviation on an international scale has resulted in all nations agreeing to abide by safety regulations developed and supervised by the International Civil Aviation Organization (ICAO). Consequently, the ICAO is responsible for a wide range of aviation safety issues including agreed standards of airworthiness, aircrew qualifications, maintenance standards in cooperation with aircraft manufacturers, limitations on flying hours and technical matters related to air traffic control. Because of the global approach to air safety regulations the number of passengers killed in aviation accidents has fallen over time while the number of passenger-kilometres, aircraft-kilometres, landings and take-offs have increased significantly (Hanlon 1996). The number of passenger deaths per annum during 1970 to 1992 fell by 75% to one death for every 1.6 billion passenger kilometre flown.

Each of the accidents identified in Table 10.1 were investigated by the governments concerned to identify the cause and determine measures to avert future events of a similar nature. Specialist air safety investigators attached to national aeronautical administrations investigate aviation accidents, report on the cause of the accident and identify measures that need to be implemented to avoid further accidents. In most cases specialist engineers from the aircraft's manufacturer assist the investigations. Where the accident was caused by a mechanical or service failure the manufacturer will quickly undertake preventative measures and advise all airlines operators of actions that are required to avert further accidents of a similar nature. Where a failure in crew training is identified, advice on amended procedures is given to all operators of the aircraft. Where there were air traffic control problems, new procedures are instituted and in some cases individuals at fault are subject to some form of penalty, either administrative of judicial. The official response to accidents involving public transport ensures that the number of accidents is minimised. Compared to the international approach adopted in aviation safety, investigation of accidents involving rail, coach and shipping transport are controlled by national governments, which may not always implement recommendations by investigating bodies.

Non-Public Transport Accidents

Compared to the level of official enquiry involving public transport, the level of official enquiry into accidents involving travellers undertaking non-public travel (walking, cycling, self drive, limited water transport) is low or does not occur except as a result of the normal administrative procedures surrounding deaths that are caused by factors other than natural causes. In most cases this may include an autopsy and an investigation by a coroner. Because incidents of this nature involve single or a small number of injuries or fatalities they rarely attract media attention. In incidents of this nature it is often difficult to apportion blame except where police investigations reveal fault either by the injured party or other parties involved in the accident. In these circumstances either civil or criminal action may follow but again the incident is unlikely to attract any level of public concern.

Some indication of the extent of the problem can be found in the number of injuries that are suffered by pedestrians. More than one-third of the 1.2 million people killed and 10 million injured globally each year in road crashes are pedestrians (Ryan 2002). Compared to the extensive research undertaken into reducing the injury and fatality levels of occupants of cars relatively little research has been directed toward reducing the incidents of injury of pedestrians.

Enhancing Transport Safety

In examining aspects of visitor safety at tourism destinations, Prideaux & Master (2001) suggested a simple four-step system for identifying visitor risk. The model (Figure 10.2) is equally suitable for accident risk identification and management of transport. Initially, potential risks are identified using databases and overseas experiences to classify injuries by severity, cause and frequency. Subsequently a profile of tourists involved in transport accidents is developed and risk management strategies are identified and implemented. Strategies are then evaluated and if deficiencies are detected the findings of the initial step are revisited via a feedback loop and the process is recommenced. The model can be used in three ways: firstly to identify patterns of behaviours that may lead to accidents; secondly to identify and implement appropriate risk management strategies; and thirdly to identify opportunities for educational programs designed to inform international visitors of potential hazards before they place themselves in a high risk situation.

Motor Vehicle Accidents Involving International Tourists in Australia

A review of injuries and accidents among international tourists in Australia and New Zealand by Page & Meyer (1997) argued for a more systematic investigation into issues

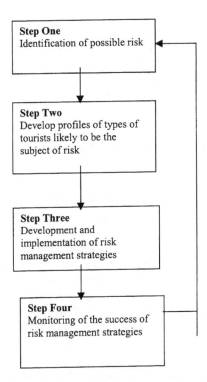

Figure 10.2: Destination risk identification and management.

Source: Prideaux & Master 2001.

of health and safety given the paucity of existing research. In a later study Wilks (1999) reported that while the medical literature identified motor vehicle accidents to be the most common cause of death of international tourists in Australia, there was little empirical research into the factors contributing to these accidents. Australia has a diversity of landscapes and climatic regions creating unique driving conditions for international visits. Unfamiliar landscapes, differences in traffic laws between Australia and other nations, long distances between major tourism areas, and lack of knowledge of specific traffic hazards that often vary according to the region travelled create a range of potentially hazardous driving conditions. Alerting international travellers to these conditions is difficult, however, to fail in addressing these issues is to put international tourists and domestic travellers sharing the roads into potentially dangerous situations.

International tourists account for approximately 2.5% of all motor vehicle fatalities in Australia, with 22.0 deaths per 100,000 persons compared with a rate of 10.8 per 100,000 population for Australian citizens. Many of the causes of accidents involving international tourists stem from a lack of familiarity with Australian road conditions, particularly in non-urban areas. According to Ellis (1999), factors contributing to accidents include fatigue (particularly in country areas), driving unfamiliar vehicles (e.g. four wheel drives and camper vans), failure to wear seat belts, alcohol, speeding, lack of knowledge of road rules and unfamiliarity with right-hand drive. Investigations into accidents involving motor vehicles driven by international visitors found that the leading factors contributing to fatalities were failure to wear seat belts (52%), alcohol (16%) and speeding (16%). Other factors previously cited may be classed as contributing factors.

Australian National Visitor Safety Program

For many years all levels of government in Australia have vigorously pursued programs to improve transport safety through enforcement, introduction of new safety standards for public transport and private vehicles, improving roads and other transport infrastructure and education. Until recently, little attention was given to the accident rate of international visitors given the comparatively small number of fatalities and the difficulty of identifying specific groups of road users out of the total population of road users. As a consequence, no particular national education or enforcement programs, apart from normal police enforcement programs and road safety campaigns targeted at domestic drivers, have been attempted prior to 2002. As the nation's tourism industry has grown in economic importance in the last two decades the authorities now recognize that in addition to the duty of care that government has to all road users there is an additional duty of care for international road users who have little practical knowledge of driving in Australia. This need for additional duty of care is a reflection of the high fatality rate of international visitors per 100,000 persons compared to domestic drivers.

When the fatality rate of international visitors is added to the fatality and injury rate from other causes, there is some potential for Australia to develop a reputation as an unsafe destination. In July 2001 all State and Territory Tourism Ministers endorsed a Queensland proposal to develop a National Visitor Safety Program designed to alert international visitors to the dangers that could be encountered during a visit to Australia. A working group was established to liase with the industry, government agencies, community organisations and individuals regarding the scope of action needed. Following the period of community and industry consultations, the working group identified the need to raise the awareness of international visitors to a number of safety issues that they might confront during their stay in Australia. These included:

- Water safety to reduce the incidence of drowning;
- Road safety;
- Safety in the bush; and
- Outback (the interior of the country) safety issues.

Some of the safety issues identified, including the bush and the outback, were unique to Australia's landscape and climate and unlikely to be encountered elsewhere.

Many of the recommendations of the Working Committee were accepted and the program was launched in April 2002 (Tourism Ministers' Council 2002). The strategy entailed a two level approach based on direct education of international visitors about safety issues, and education of tourism operators about visitor safety issues and their associated responsibilities. Initiatives introduced under the direct education program included:

• Visitor safety information on key tourism web sites;
• Safety information in visitor publications distributed in the visitor's home country and in Australia;
• Production of a video tape to be shown on airlines flying into Australia;
• Production of a multilingual booklet titled Safety Tips for Visitors to be distributed at key tourist locations; and
• Visitor safety information in in-flight magazines.

Initiatives adopted to encourage tourism operators to become active in visitor education included:

• Safety issues in accreditation programs;
• Working with key industry organisations to educate them about visitor safety issues and to encourage them to inform visitors about safety issues; and
• Enhanced linkages with government agencies, community organisations and the tourism industry to ensure that safety information is distributed through appropriate channels.

The effectiveness of the National Visitor Safety Program will not be apparent for some time and requires careful evaluation to identify deficiencies in the current suite of program initiatives. In terms of the model outlined in Figure 10.2, the National Visitors Safety Program has broadly followed Steps One to Three. However, Step Four needs to be included to evaluate its success if it is to achieve the long-term goal of reducing visitor injuries.

Conclusions

The authorities often ignore tourism safety. The fears raised by the sight of aircraft crashing into New York's World Trade Center testifies to the reaction that tourists exhibit when they lose faith in the ability of either a destination or a transport mode to ensure their personal safety. Public transport accidents receive considerable media attention and create public concern forcing governments to react by establishing and maintaining a range of safety organisations. Conversely, accidents involving non-public transport receive little attention yet are associated with a higher rate of fatalities and injuries amongst international tourists.

The adoption of the model suggested in Figure 10.2 as a means of identifying transport risks and implementing preventative strategies is important if destinations are to discharge their duty of care towards visitors. Unfortunately, many destinations are yet to accept their responsibilities in this regard. To fail to incorporate and maintain a high level of safety in the national transport system will eventually place a destination in an unfavourable competitive position compared to others that show they value visitors safety through the implementation of safety programs such as the National Visitor Safety Program recently introduced in Australia.

Given the potential for transport injury deaths to affect destination choice and the lack of existing research into this aspect of consumer choice, there is an urgent need to undertake further investigation into transport safety as a factor in destination choice. It can be expected that research of this nature will assist planners to quantify the impact of transport disasters on tourism demand.

References

Cartwright, R. (2000). Reducing the health risks associated with travel. *Tourism Economics, 6* (2), 159–167.

Chung-yan, C. (2002). 12 Tragic episodes that have left hundreds dead. *Sunday Morning Post,* (Hong Kong), 2.

Ellis, B. (1999). A national road safety overview. In: J. Wilks, B. Watson, & R. Hansen (Eds), *International visitors and road safety in Australia: A status report* (pp. 19–26). Canberra: Australian Transport Safety Bureau.

Hanlon, P. (1996). *Global airlines competition in a transnational industry.* Oxford: Butterworth Heinemann.

Hunter-Jones, J. (2000). Identifying the responsibility for risk at tourism destinations: The U.K. experience. *Tourism Economics, 6* (2), 187–198.

Mawby, A. (2000). Tourist's perceptions of security: The risk-fear paradox. *Tourism Economics, 6* (2), 109–122.

May, V. (1989). Tourism health taking action. *Tourism Management, 10* (4), 341.

Page, S. J., & Meyer, D. (1997). Injuries and accidents among international tourists in Australasia: Scale, causes and solutions. In: S. Clift, & P. Grabowski (Eds), *Tourism and health: risks, research and responses* (pp. 61–79). London: Pinter.

Prideaux, B. (2000). The resort development spectrum. *Tourism management, 21* (3), 225–241.

Prideaux, B., & Master, H. (2001). Reducing risk factors for international visitors in destinations. *Asia Pacific Journal of Tourism Research, 6* (2), 24–32.

Ryan, S. (2002). Pedestrians drive safety push. *The Courtier Mail,* (13 May), 6.

Sonmez, S. F., & Graefe, A. R. (1998). Influence of terrorism risk on foreign tourism decisions. *Annals of Tourism Research, 25* (1), 112–144.

Tak-ho, F. (2002). Fresh blow to already tarnished reputation, *Sunday Morning Post,* (Hong Kong), 2.

Tourism Ministers' Council (2002). *Safety tips for visitors to Australia.* Brisbane: Tourism Queensland (booklet and video).

Tsaur, S., Tzeng, G., & Wang, K. (1997). Evaluating tourist risks from fuzzy perspectives. *Annals of Tourism Research, 24* (4), 796–812.

Turner, P., Delahunty, B., Greenway, P., Lyon, J., Taylor, C., & Willett, D. (1997). *Indonesia.* Hawthorn: Lonely Planet.

Watson, B., Wilks, J., & Hansen, R. (1999). Critical issues and future directions. In: J. Wilks, B. Watson, & R. Hansen (Eds), *International visitors and road safety in Australia: A status report* (pp. 149–159). Canberra: Australian Transport Safety Bureau.

Wilks, J. (1999). International tourists, motor vehicles and road safety: A review of the literature leading up to the Sydney 2000 Olympics. *Journal of Travel Medicine, 6,* 115–121.

Wilks, J., Watson, B., & Hansen, R. (Eds) (1999). *International visitors and road safety in Australia: A status report.* Canberra: Australian Transport Safety Bureau.

Chapter 11

Towards a Framework for Tourism Disaster Management*

Bill Faulkner

Introduction

To the casual observer exposed to the plethora of media that currently inform our daily lives, it appears that we live in an increasingly disaster prone world. This perception has some foundations, at least to the extent that the number of disasters (defined in terms of declarations of disaster areas, economic value of losses and the number of victims) has, in fact, increased in recent decades (Blaikie *et al*. 1994). However, the same authors point out that the incidence of natural hazard events (earthquakes, eruptions, floods or cyclones) has not increased, while others have suggested that the definition of disasters has become too fluid for statistical time series purposes (Horlick-Jones *et al*. 1991). Notwithstanding statistical uncertainties, there is a body of opinion, which has attributed the apparent increase in the human toll of disasters to a combination of population growth, increased urbanisation and global economic pressures (Blaikie *et al*. 1994; Berke 1998; Brammer 1990; Burton *et al*. 1978; Donohue 1982; Hartmann & Standing 1989). In particular, it is suggested that these factors have either resulted in human settlement and activity being extended into areas which have increased exposure to hazards, or these activities have actually been instrumental in inducing hazards.

In observing that our environment appears to have become increasingly 'turbulent and crisis prone', Richardson (1994) has suggested this might be so not only because we have become a more crowded world, but also because we now have more powerful technology that has the capacity to generate disasters. As the spectre of the Millenium Bug illustrates, for instance, computer failures can bring major computer-driven systems to a standstill instantaneously. The complexity of technology-based systems means that they are more prone to the 'butterfly effect' described by Edward Lorenz (1993) and presented as one of a centrepiece of chaos theory (Gleick 1987). Small changes or

---

* This chapter first appeared in 2001 in *Tourism Management*, 22 (1), 135–147.

Managing Tourist Health and Safety in the New Millennium
© 2003 Published by Elsevier Science Ltd.
ISBN: 0-08-044000-2

failures in the system can precipitate major displacement through mutually reinforcing positive feed back processes. Mitroff (1988) has alluded to this in his reference to the role of the interaction between information technology and economic systems in creating wild swings in the financial system. The role of technology in exposing humankind to 'natural' disasters is succinctly described in the following remarks by Burton *et al.* (1978: 1–2):

> In a time of extraordinary human effort to control the natural world, the global toll from extreme events of nature is increasing . . . It may well be that the ways in which mankind deploys its resources and technology in attempts to cope with extreme events of nature are inducing greater rather than less damage and that the process of rapid social change work in their own way to place more people at risk and make them more vulnerable.
> . . . To sum up, the global toll of natural disaster rises at least as fast as the increase in population and material wealth, and probably faster.

Whether the incidence of disasters is increasing, or it is simply a matter of each disaster having more devastating effects, as the above summary suggests, it is apparent we live in an increasingly complex world and this has contributed to making us more crisis or disaster prone (Richardson 1994). Complexity, in this context, refers to an intricacy and coherence of natural and human systems, which complicates the process of isolating cause and effect relationships in the manner so often assumed as being possible in traditional research. For this reason, the boundaries between natural disasters and those induced by human action are becoming increasingly blurred, and this element of disaster situations needs to be taken into account in any analysis of such phenomena (Capra 1996; Waldrop 1992). As an area of human activity, tourism is no less prone to disasters than any other. Indeed, it has been suggested that the increased volume of global tourism activity has combined with the attractiveness of high-risk exotic destinations to expose tourists to greater levels of risk (Drabek 1995; Murphy & Bayley 1989). Despite this, relatively little systematic research has been carried out on disaster phenomena in tourism, the impacts of such events on the tourism industry and the responses of industry and relevant government agencies to cope with these impacts. Such research is an essential foundation for assisting the tourism industry and relevant government agencies to learn from past experiences, and develop strategies for avoiding and coping with similar events in the future.

One of the reasons so little progress has been made in the advancing of our understanding of tourism disasters is the limited development of the theoretical and conceptual frameworks required to underpin the analysis of this phenomenon. The purpose of the current study is to fill this gap, by using the broader literature relating to crises and disaster management as a foundation for such a framework. The first step in this process involves the establishment of a distinction between crises and a disaster, which goes some way towards clarifying the complexity issue, alluded to above. Community (and organisational) responses to disaster situations are then examined with a view to providing some insights into the essential ingredients of disaster management strategies. Finally, aspects of tourism disasters are examined as a step towards providing a model for developing tourism-specific disaster management strategies.

The Nature of Disasters and Crises

Much early management theory assumed relative stability in both internal and external environments of organisations and, therefore, did not provide a firm foundation for coping with change and crises (Booth 1993). If the implications of change were considered at all, this was viewed in terms of the challenge of coping with gradual (relatively predictable) change, rather than sudden changes which might test the organisation's ability to cope. Such situations might be described as crises or disasters, depending on the distinctions referred to below. Given the specific focus of this paper on tourism sector adjustments to disasters at the destination level, it should be borne in mind that references to 'organisations' in this section apply equally to destinations and host communities.

One perspective on the nature of crises is provided by Selbst (1978), who refers to a crisis as, "Any action or failure to act that interferes with an (organisation's) ongoing functions, the acceptable attainment of its objectives, its viability or survival, or that has a detrimental personal effect as perceived by the majority of its employees, clients or constituents". There are two dimensions of the crisis situation emphasised in this definition, which shed light on the distinction between crises and disasters, and the ramifications of these two situations with regard to the responses of organisations and communities. Firstly, by referring to 'any action or failure to act', Selbst implies that the event in question is in some way attributable to the organisation itself. Secondly, it is implied that the event must have detrimental or negative effects on the organisation as a whole, or individuals within it.

Selbst's definition of crises seems to exclude situations where the survival of an organisation or community is placed in jeopardy because of events over which those involved have little or no control. For example, tornadoes, floods and earthquakes can hardly be regarded as self-induced, although communities in vulnerable areas can take steps to minimise the impacts of such events. Thus, for the purposes of this analysis, it is proposed that 'crisis' be used to describe a situation where the root cause of an event is, to some extent, self-inflicted through such problems as inept management structures and practices or a failure to adapt to change. On the other hand, 'disaster' will be used to refer to situations where an enterprise (or collection of enterprises in the case of a tourist destination) is confronted with sudden unpredictable catastrophic changes over which it has little control. We can therefore envisage a spectrum of events such as that depicted in Figure 11.1, with crises located at one extreme and disasters at the other. However, as implied in the introduction, it is not always clear where we locate specific events along this continuum because, even in the case of natural disasters, the damage experienced is often partially attributable to human action.

Good management can avoid crises to some degree, but must equally incorporate strategies for coping with the unexpected event over which the organisation has little control. Frequently, the recognition of a critical problem that might eventually precipitate a crisis becomes a matter of too little too late largely because, as Booth (1993: 106) observes, 'standard procedures tend to block out or try to redefine the abnormal as normal.' This problem is probably more relevant to the genesis of crises, where organisations fail to adapt to gradual change, but it might apply to disaster

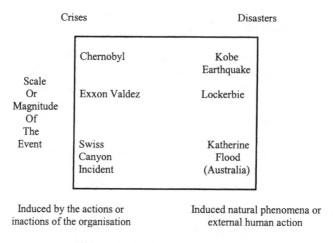

Figure 11.1: Crises and disasters.

situations to the extent that the tendency to ignore warnings of an impending disaster often leaves communities unprepared when it actually happens.

Crises and disasters epitomise chaos phenomena as it is described by such authors as Gleick (1987), Peat (1991), Prigogine & Stengers (1985) and Faulkner & Russell (1997) in the tourism context. In terms of Chaos Theory, even apparently stable systems are frequently 'at the edge of chaos', whereby a seemingly insignificant event may be enough to precipitate instability and change on such a scale that the integrity and coherence of the system appears to be threatened. Fink emphasises the ubiquity of the 'edge of chaos' condition in business when he suggested that businesses generally are a crisis waiting to happen — i.e. "any time you (i.e. managers) are not in crisis, you are instead in a pre-crisis, or prodromal mode" (Fink 1986: 7). In his view, the essence of crisis management thus becomes "the art of removing much of the risk and uncertainty to allow you to achieve more control over your destiny" (Fink 1986: 15).

The other aspect of Selbst's definition highlighted above concerns its emphasis on the negative and threatening impacts of the event, rather than the possibility of it representing a turning point or opportunity. As Fink (1986) emphasises, the Webster dictionary definition of a crisis refers to such events as "a turning point for better or worse". Crises and disasters therefore have transformational connotations, with each such event having potential positive (e.g. stimulus to innovation, recognition of new markets, etc.), as well as negative outcomes. This is illustrated by seasonal floods in riverine areas of Peninsula Malaysia, which are seen as both hazards and resources (Chan 1995). The floods bring disruption to communities within the area, but at the same time they replenish the productive capacity of riverine alluvial soils upon which the region's agricultural industry is so dependent.

Again, this is consistent with elements of Chaos Theory, which see chaos as essentially creative, rather than a destructive process. Once a system is pushed past some point of criticality by some crisis or disaster, it may well be destroyed as an entity, it might be restored to a configuration resembling its pre-crisis/disaster state, or a totally new and more effective configuration might emerge. The potential for both destructive and positively creative forces being unleashed by the chaos associated with crises and disasters is illustrated in Berman & Roel's (1993: 82) description of reactions to the 1985 Mexico City earthquake:

> Crises bring about marked regressions as well as opportunities for creativity and new options. They are turning points in which regressive tendencies uncover discrimination (and) resentment about ethnic and socioeconomic differences . . . yet they also trigger progressive potentials and solidarity.

The above discussion has provided an explanation of the distinction being drawn between crises and disasters for the purposes of the current study, along with a description of the general characteristics of these phenomena. However, we are no closer to precisely defining crises and disasters in a form which would enable us to empirically identify when such situations occur. In turning to this issue, it is important that we be reminded that, from an organisational point of view, crises and disasters are essentially very similar and the main distinction between them is a root cause of the problem. The former represent situations where the causes of the problem are associated with on-going change and the failure of organisations to adapt to this, while the latter are triggered by sudden events over which the organisation has relatively little control. Notwithstanding this distinction, most of the features attributed to disasters in the following discussion are equally applicable to crises.

Carter (1991: xxiii) defines a disaster as "an event, natural or man-made, sudden or progressive, which impacts with such severity that the affected community has to respond by taking exceptional measures". In his definition of crises, Booth (1993) places a similar emphasis on the necessity of 'exceptional measures' in the community's response by referring to the necessity of non-routine responses, but he adds that stress is created by the suddenness of the change and the pressure it places on adaptive capabilities. Thus, a crisis is described as "a situation faced by an individual, group or organisation which they are unable to cope with by the use of normal routine procedures and in which stress is created by sudden change" (Booth 1993: 86).

Several other authors have attempted to distill the essential characteristics of disaster or crisis situations (Fink 1986: 20; Keown-McMullan 1997: 9; Weiner & Kahn 1972: 21). A synthesis of these contributions produces the following key ingredients:

- A triggering event, which is so significant that it challenges the existing structure, routine operations or survival of the organisation;
- High threat, short decision time and an element of surprise and urgency;
- A perception of an inability to cope among those directly affected;
- A turning point, when decisive change, which may have both positive and negative connotations, is imminent. As Keown-McMullan (1997: 9) emphasises, 'even if the crisis is successfully managed, the organisation will have undergone significant change';
- Characterised by 'fluid, unstable, dynamic' situations (Fink 1986: 20).

Another approach to defining disasters is provided by Keller and Al-Madhari (1996: 20), who applied arbitrary statistical benchmarks. Thus, disasters were defined in terms of a threshold number of fatalities (10), damage costs (US$1 million) and number of people evacuated (50). On the basis of this definition, they claim there have been 6,000 disasters since 1970, with 4 million deaths and widespread economic costs. This approach has the appeal of providing a solid, unambiguous foundation for defining disasters, and it is appropriate in the context of studies concerned about statistical issues, such as probabilistic prediction of frequency and magnitude of disasters. However, it loses sight of the qualitative factors referred to above, which are present in disaster situations irrespective of whether or not the fatality, damage cost and evacuation thresholds are reached.

Community Responses and the Ingredients of Disaster Recovery Strategies

The reference to both crises and disasters, and the subtleties of the distinction between them, has been useful for the purpose of highlighting how disasters, or at least the severity of their impacts, are to varying degrees influenced by the actions of the individuals, organisations or communities that are affected by them. This section focuses on disasters specifically. That is, those situations where the event which disrupts the routine of the community concerned, and in response to which adjustments have to be made, is triggered externally. We are interested in how communities and individuals respond to these events, and the implications of this for the development of disaster strategies.

From a sociological perspective, the immediate response to a disaster situation has been observed as including several phases (based on Arnold (1980) in Booth (1993: 102–103)).

- Shock at both the individual and the collective level, where the unexpected nature of the event and the severity of its impacts cause stress and a sense of helplessness and disorientation. While the stressfulness of the situation may initially impair adaptive responses, it is also a mobilising factor for those involved;
- Denial or defensive retreat. Denial being an attempt to reach back to the safety of the known, or an attempt to avoid the crisis by repressing it. Defensive retreat may involve either evacuation from the effected area, or a strategic withdrawal to safe places within the area. Evasive action is taken to ensure safety and this enables those concerned to regroup;
- An acknowledgement represents a turning point whereby the community accepts the reality of the change; and
- Adaptation, where the community learns from the crisis, develops new ways of coping and rebuilds.

This sequence is probably as much applicable to the individual level as it is at the collective (i.e. organisational and community) level. Chan (1995) suggests that, beyond the immediate occurrence of the disaster, responses might take one of several broader courses:

- To protect (prevent or modify disasters);
- To accommodate (change human use systems to suit disasters);
- To retreat (resettlement elsewhere);
- To do nothing.

It is hard to envisage situations where the 'do nothing' strategy provides a viable alternative. Even in the case of the random, one-off event, which has a very low probability of recurrence, some sort of recovery plan needs to be implemented so that lingering longer-term impacts of the disaster can be ameliorated. The applicability of the remaining three strategies will clearly depend on the extent of damage caused by the disaster, the probability and frequency of recurrence and the adaptability of the impacted community.

At some point in a disaster situation there needs to be an assessment of the capacity of the community to cope so that the appropriate level of emergency relief can be determined. A more than appropriate level of relief involves wastage and the unnecessary straining of resources, while insufficient external support will exacerbate the effects of the disaster. However, according to Granot (1995), there are few objective measures for assessing a community's resilience in this regard. Geipel (1982) ranked levels of community impact in a manner that implies a continuum between minimal effects and total collapse. The following categories were therefore identified:

- communities which have not suffered and therefore have the capacity to support others which have;
- communities which escape with only limited loss of life and property — community systems remain largely intact and normal built-in elasticity of resources permits self-recovery;
- communities that sustain so much damage that they can only recover with outside help. With such help their own systems are capable of coping and ultimate recovery;
- communities that are devastated so much that community systems collapse.

Granot challenges the notion of a continuum, however, by suggesting that the effect of a disaster on a community might be more appropriately represented in terms of the percolation principle, which sees changing states as being non-linear. He suggests that, "once a certain threshold is crossed in a sufficient number of constituent subsystems, a basic change takes place in the community system as a whole, affecting its overall capacity to cope" (Granot 1995: 6). It is suggested that factors affecting a community's capacity to cope include:

- Community background factors (relevant demographic, socioeconomic, political, cultural, organisational and resource level characteristics);
- Event factors (objective factors precipitating the cause or causes of the incident); and
- Impact factors (immediate discernible outcome, as reflected in such factors as number of casualties, property damage, etc.).

Richardson's (1994) analysis of crisis management in organisations provides another perspective on community adjustment capabilities by distinguishing between 'single' and 'double loop' learning approaches. In the former, the response to disasters involves a reorientation 'more or less in keeping with traditional objectives and traditional roles' (Richardson 1994: 5). Alternatively, the 'double loop' learning approach challenges traditional beliefs about "what society and management is and should do". This approach recognises that management systems in place can themselves engender the ingredients of chaos and catastrophe, and "organisations must be prepared to manage through the crisis driven era that is, in one sense, given to them but managers must also be more aware and proactively concerned about organisations as the creators of crises" (p. 6).

For the purposes of exploring these issues further and examining the ingredients of disaster management strategies in more detail, it is useful to look at the frameworks that have been used to describe the stages in response to disasters at the community level. Two such frameworks have been produced, one by Fink (1986) and the other by Roberts (1994). These are described in Table 11.1, where a composite set of stages drawing upon both frameworks is also presented in the first column. The latter set will be utilised as the basis for further discussion because it is more comprehensive.

Community responses to disasters, both during the emergency and afterwards in the recovery period, involve many different organisations (Granot 1997; Huque 1998). In this situation, it is not uncommon for competition and rivalry among these organisations to become a major impediment to both coordination and the ability of organisations to respond effectively (Comfort 1990). In their examination of emergency services responses to the 1989 Newcastle earthquake, for instance, Kouzmin *et al.* (1995) have noted that ambiguities in the division of responsibilities and the rivalries between various emergency service agencies undermined the effectiveness of their response collectively. These problems arise from, and are intensified by, the scarcity of public resources and the necessity of each organisation to justify its existence in order to obtain a share of these resources.

In addition to this problem, different internal cultures and modus operandi become barriers to communication and co-operation between organisations. As Granot (1997: 309–310) has observed, "Old jurisdictional disputes can often be set aside and left unresolved in normal times, but in emergencies they have a way of returning as conflicts that prevent coordination" and as "these situations are difficult to resolve in the heat of the moment, it is clear inter-organisational relationships need to be planned ahead and exercised before the actual need occurs". Quarantelli (1982) also sees this as a major challenge in the disaster preparation. However, while coordination between emergency services in the development and implementation of disaster strategies is given lip service, it is seldom reflected in actions (Hills 1994). On the other hand, Granot (1997) also suggests that inadequate resources often force agencies into a collaborative arrangement under emergency conditions. The extent to which pressure to react to the disaster might force organisations to work together, and thus provide a catalyst for breaking down institutional barriers in the longer term, is not clearly addressed in the literature.

Table 11.1: Stages in a community's response to a disaster.

Composite stages	Fink's (1986) stages	Robert's (1994) stages
1. Pre-event		*Pre-event*: where action can be taken to prevent disasters (e.g. growth management planning or plans aimed at mitigating the effects of potential disasters)
2. Prodromal	*Prodromal stage*: when it becomes apparent that the crisis is inevitable	
3. Emergency	*Acute stage*: the point of no return when the crisis has hit and damage limitation is the main objective	*Emergency phase*: when the effects of the disaster has been felt and action has to be taken to rescue people and property
4. Intermediate		*Intermediate phase*: when the short-term needs of the people affected must be dealt with – restoring utilities and essential services. The objective at this point being to restore the community to normality as quickly as possible
5. Long term (recovery)	*Chronic stage*: clean-up, post-mortem, self-analysis and healing	*Long-term phase*: continuation of the previous phase, but items that could not be addressed quickly are attended to at this point (repair of damaged infrastructure, correcting environmental problems, couselling victims, reinvestment strategies, debriefings to provide input to revisions of disaster strategies)
6. Resolution	*Resolution*: routine restored or new improved state	

The degree to which emergency services and other organisations can be prepared for disasters is questioned by Huque (1998), who notes that policies and decision-making structures that govern an organisation's activities in normal times may not be appropriate in disaster situations. For example, the hierarchical structure and chain of command under normal conditions is necessary for internal coordination purposes, but the tight time lines for decisions and actions during emergencies may make this structure too unresponsive. Bureaucratic structures and power relationships restrict the ability of organisations to respond promptly and effectively to emergency conditions, and this in itself constitutes a barrier to inter-agency cooperation. Heath (1995) observed how response times of emergency services and government agencies in the case of the 1995 Kobe earthquake were affected by bureaucratic procedures. In this case Japanese cultural orientation toward bottom-up consensus in decision-making also affected the timeliness of the response.

Other factors cited by Heath (1995) as impeding the timeliness and effectiveness of the response in the Kobe case provide some useful insights into some of the key considerations in the development of disaster plans. These included:

- Communication failures;
- Availability of resources (a common problem as governments and response agencies rarely set aside resources in reserve for infrequent and unpredictable crises);
- Deployment of resources at a distance from the impact area can be slow — aggravated by damage to infrastructure;
- The attention and efforts of many affected by the disaster shifts from the big picture to immediate, more local concerns and they resist efforts to redirect their attempts towards more coordinated action. Dispersion of resources becomes a related problem;
- "Even without blocked roads and dispersed response demands, a large magnitude impact will create demands for service that exceeds the capabilities of response agencies" (Heath 1995: 17).

Also, on the basis of his analysis of responses to the Kobe incident, Heath has emphasised the need to incorporate a 'cascaded strategic priority profile' (CSPP), in the disaster planning phase. This approach involves "a rank ordering of tasks and activities that need to be undertaken, moving from the highest to lowest priority" (Heath 1995: 18). CSPPs need to be developed at various levels of managerial operation, from overall to the local level, to provide multiple layers. Furthermore, these must be articulated with each other to avoid waste, duplication and mutually antagonistic actions.

The role of media in disaster management strategies can be crucial to such an extent that it might make the difference between whether or not a difficult situation evolves into a disaster (Fink 1986; Keown-McMullan 1997). Media outlets can help by disseminating warnings in the lead up to impending disasters (i.e. where these are predictable) and providing information during the recovery stage. However, they can also hinder emergency operations by spreading false information or criticising these operations in a manner that distracts the emergency service personnel from their task. Thus Fischer & Harr (1994) found that up to 20% of emergency operating centre staff's time in the Andover, Kansas tornado incident was spent on media damage control. Furthermore, adverse media reports had an impact on the decisions made by these staff. In Australia, Christine (1995) has noted the effect of misleading reports exaggerating the extent of the 1993 Sydney bushfires. These experiences highlight the necessity of a media communication strategy involving the early establishment of a centralised source in order to ensure that misleading and contradictory information is not disseminated (Riley & Meadows 1997).

On the role of the media in disasters in the United States, Quarantelli (1996) has observed:

- Disaster preparedness planning in mass media is generally poor;
- Some of the coordination problems attributed to emergency services agencies are also evident in the press as "local mass media systems consider disasters in their own community as 'their' disasters" and "this is sometimes manifest in tension . . . between local mass media and national network staff members" (Quarantelli 1996: 6);
- There is a tendency for selective reporting focusing on the activities of formal organisations (with whom the media has established links), rather than emergent and informally organised volunteers;
- Television, in particular, is prone to perpetuate disaster myths, such as the persistence of disruption, panic, looting, etc.

Disaster strategies clearly need to articulate a set of appropriate actions for each of the stages described in Table 11.1. Quarantelli (1984) and Turner (1994) have each commented on the essential ingredients of disaster strategies from two different, but complementary, perspectives. As reflected in Table 11.2, while the latter provides useful guidelines for the production of the strategy, Quarantelli concentrates more on its actual ingredients and the systems that need to be in place to make it work. By amalgamating the analyses of both authors, therefore, it is possible to produce a more comprehensive structure for producing a disaster survival strategy.

From this brief overview of community responses to disasters, and on the basis of other insights provided by analyses of disaster situations, a number of conclusions can be drawn about important considerations in disaster preparedness generally. These will be revisited in the next section after aspects of tourism disasters have been explored, so that they can be related specifically to tourism disaster strategy development considerations.

Tourism Disasters

Several authors have emphasised the vulnerability of tourist destinations, and thus tourists, to disasters and some have suggested that, in these situations, tourists might be more exposed to danger than anyone else (Drabek 1995). Murphy & Bayley (1989) suggest that the exposure of tourism to natural disasters is linked with the attractiveness of many high-risk exotic locations, where events such as hurricanes, avalanches and volcanic activity are common. They are also at risk from hijacking and terrorism because, as Lehrman (1986) observes, tourists have become soft targets in a period when increased security measures have made traditional targets (politicians, embassies, etc.) less attractive for terrorists. Furthermore, tourists themselves are often more vulnerable than locals in disaster situations because they are less familiar with local hazards and the resources that can be relied on to avoid risk, and they are less independent (Burby & Wagner 1996; Drabek 1992; 1994).

Table 11.2: Ingredients of a disaster survival strategy.

Strategy development (Turner 1994)	Implementation (Quarantelli 1984)
• Form disaster recovery committee and convening meetings for the purpose of sharing information • Risk assessment (identify potential threats/disasters and prioritise in terms of probability of occurrence – real, likely and historical threats. Perhaps stimulated by a definition and classification of potential disasters) • Analysis of anticipated short- and long-term impacts • Identification of strategies for avoiding/minimising impacts, critical actions necessary, chain of command for coordination, responsibilities and resources • Prepare and disseminate manual and secure commitment from responsible parties and relevant agencies. Relevant contact information must be included	• Holding disaster drills, rehearsals and simulations • Developing techniques for training, knowledge transfer and assessments • Formulating memoranda of understanding and mutual aid agreements • Educating the public and others involved in the planning process • Obtaining, positioning and maintaining relevant material resources • Undertaking public educational activities • Establishing informal linkages between involved groups • Thinking and communicating information about future dangers and hazards • Drawing up organisational disaster plans and integrating them with overall community mass emergency plans • Continually updating obsolete materials/strategies

Despite the potentially devastating effect natural and man-made disasters can have on tourism, few tourism organisations at the enterprise or destination level have properly developed disaster strategies as an integral part of their business plans (Cassedy 1991). In studies of disaster preparedness among tourism industry enterprises in the United States, Drabek (1992; 1995) has reported that, while there was a relatively high degree of disaster preparedness among tourism executives, this was qualified by the observation that many had essentially informal (undocumented) strategies in place and these strategies addressed only one type of hazard. Also, the level of staff turnover had not been taken sufficiently into account in the consideration of the frequency of staff education and some misconceptions about disaster effects (e.g. inflated expectations regarding the potential for looting) influenced the planned response. Furthermore, larger firms with more professional senior managers and planning resources tended to be more prepared than the many smaller establishments. Elsewhere, Burby & Wagner (1996) reported a high degree of preparedness among hotel establishments in New Orleans, but this preparedness was compromised by similar reservations as those raised by Drabek.

The critical role of the media in disaster situations has been referred to in the previous section. In tourism context, the impacts of disasters on the market are often out of proportion with their actual disruptive effects because of exaggeration by the media (Cassedy 1991; Murphy & Bayley 1989; Drabek 1992). As Young & Montgomery (1998: 4) have observed, ". . . a crisis has the potential to be detrimental to the marketability of any tourist destination, particularly if it is dramatised and distorted through rumours and the media". Meanwhile, disaster situations provide a fertile ground for misinformation, as disruptions to communications systems combine with publication deadlines to inhibit the verification of reports and the ratings game fosters sensationalism (Milo & Yoder 1991).

Media reports have the potential to have a devastating impact on disaster-affected destinations because pleasure travel is a discretionary item and, within the mind of the consumer, the "quest for paradise (can) suddenly transform into a dangerous journey that most travellers would rather avoid" (Cassedy 1991: 4). This effect is expressed in econometric terms by Gonzalez-Herrero & Pratt (1998: 86) when they suggest that "tourism demand presents a higher elasticity index per level of perceived risk than any other industry because of the hedonistic . . . benefits customers ascribe to its products and services". By virtue of the power of the media and the tendency for negative images to linger, the recovery of destinations usually takes longer than the period required for the restoration of services to normalcy. This has been observed in a number of case studies, including the 1987 Fiji Coup, the 1989 San Francisco Earthquake, the 1989 Tiananmen Square incident (Cassedy 1991) and Hokkaido's Mount Usu volcanic eruption (Hirose 1982). The effectiveness with which the tourism industry in a disaster area handles a crisis, and therefore the degree to which it is prepared for it, has a bearing on how quickly services are restored to normal. However, the speed of the destination's recovery ultimately hinges on the degree to which market communication plans have been integrated with disaster management strategies.

The above observation perhaps explains why some models for tourism disaster plans (see, for example, Young & Montgomery 1998) tend to emphasise market communication considerations at the expense of other aspects. This approach, however, involves the risk of a counter-productive over-reaction, which is illustrated by Cammisa's (1993) observations regarding the response of Florida tourism authorities to Hurricane Andrew in 1992. Cammisa suggested that, in their eagerness to assure the market that Florida's hotel accommodation had not been affected by the hurricane, tourism authorities overlooked the fact that the rest of the tourism infrastructure in the area had been devastated. Thus, he pointed out that, 'rather than face up to this reality, unfortunately 'denial' communication emanated from the area' (Cammisa 1993: 294). A similar denial pattern was observed in the response of authorities to the Miami tourist crime incidents in 1992 and 1993. Another form of denial is noted by Murphy and Bayley (1989: 38) who highlight the reticence of tourism operators to bring attention to hazards and the need to take precautions, yet "safety drills and messages have become standard features of sea and air travel".

Bearing in mind the distinction drawn between crises and disasters earlier in this paper, many disasters are not predictable and their disruptive effects are generally unavoidable. However, through the development of a disaster management strategy, many potential hazards can either be totally avoided, or at least their impacts can be minimised as a consequence of the prompt responses facilitated by the plan. Furthermore, confusion and the duplication of effort can be avoided, leading to a more efficient response, while the establishment of a preset plan to guide the responses of those involved results in the potential for panic and stress being reduced (Cassedy 1991).

As mentioned previously, Young & Montgomery (1998) have provided a model for a detailed crisis management plan, although this tends to emphasise communication aspects. Other more balanced models have been developed by Cassedy (1991) and Drabek (1995). The main ingredients of each of the latter models are identified in Table 11.3, which reveals contrasting orientations in these two contributions. Cassedy emphasises aspects of the process of developing effective strategies, while Drabek's approach is structured around the sequence of responses that is necessary to cope with the emergency. While each of these authors provides a useful contribution, the cross-fertilisation between their respective contributions has been limited and few insights have been drawn from the substantial literature on disaster and crises management responses referred to in the earlier sections of this chapter.

Another approach to analysing responses to crises is based on operations research, whereby linear programming techniques are used to identify optimal responses (Arbel

Table 11.3: Ingredients of tourism disaster strategies.

Cassedy (1991)	**Drabek (1995)**
• *Selection of a team leader*: a senior person with authority and able to command respect (ability to communicate effectively, prioritise and manage multiple tasks, ability to delegate, coordinate and control, work cohesively with a crisis management team, make good decisions quickly) • *Team development*: a permanent and integral feature of strategic planning; able to identify and analyse possible crises, develop contingency plans • *Contingency plan*: including mechanism for activating the plan, possible crisis, objectives, worst-case scenario, trigger mechanism • *Actions*: action plan assignment of tasks, including gathering information and developing relationships with other agencies/groups (govt. agencies, other travel providers, emergency services, health services, the media, the community, the travelling public) • *Crisis management command centre*: a specific location and facility with relevant communication and other resources for the crisis management team	• Warning • Confirmation • Mobilisation • Customer information • Customer shelters • Employee concerns • Transportation • Employee sheltering • Looting protection • Re-entry issues

& Bargur 1980). While this approach has been effective in systems where the parameters can be tightly specified in quantitative terms (e.g. a manufacturing organisation and individual hotel chain operations), it would appear to be less applicable to a tourist destination. A tourist destination encompasses more loosely connected systems (social, economic, environmental, physical infrastructure) than in the case of a single firm and responses to externally induced shocks must take into account the more fluid relationships between the various parties concerned.

Both the linear programming approach and, to a lesser extent, the other approaches described so far implicitly assume that the events creating the crisis situation are invariably temporary aberrations, and that the primary objective is to restore the system to the pre-existing (pre-shock) equilibrium. However, insights from Chaos Theory perspective described earlier in this paper suggest that this view of crises and disasters may be deficient in two respects. Firstly, some shocks have lingering effects that make the pre-shock equilibrium a redundant (or at least sub-optimal) approach with regard to longer-term sustainability. That is, in terms of the chaos framework, systems are often 'at the edge of chaos' and a single event can set in train a series of positive feedback loops (or chain reaction) which make the pre-event status-quo no longer viable. Secondly, the chaos created by crises can be a creative process, with the potential for innovative new configurations emerging from the 'ruins'. In this sense, crisis can act as a trigger or catalyst for a more vigorous and adaptable tourism industry at a destination. This effect is alluded to in Murphy & Bayley's (1989) reference to the aftermath of the Mount St Helens eruption. Here, recovery measures put in place after the emergency led to additional resources being devoted to tourism development in the affected area, and thus an overall improvement over pre-disaster conditions. The site of the disaster became an attraction in its own right.

Insights derived from the general analysis of disaster and crises management strategies in earlier sections have been combined with those obtained from the more specific examination of tourism disaster strategies in this section to produce a generic framework for tourism disaster strategies in Table 11.4. The details of this table have been largely dealt with in the preceding discussion. However, the underlying rationale of the framework hinges on several fundamental principles, which warrant some emphasis at this point because they summarise the main conclusions emerging from the study and highlight important implications for future research. These principles are outlined below in terms of the prerequisites for, and ingredients of, effective disaster management strategies.

Prerequisites of effective tourism disaster management planning include:

- *Coordinated, team approach.* Given the range of private and public sector organisations that are directly and indirectly involved in the delivery of services to tourists, the development and implementation of a tourism disaster strategy requires a coordinated approach, with a designated tourism disaster management team being established to ensure that this happens. This team needs to work in conjunction with various other public sector planning agencies and providers of emergency services in order to ensure that the tourism industry's action plan dovetails with that of these other parties.

Table 11.4: Tourism disaster management framework.

Phase in disaster process	Elements of the disaster management responses	Principal ingredients of the disaster management strategies
1. Pre-event When action can be taken to prevent or mitigate the effects of potential disasters	*Precursors* • Appoint a disaster management team (DMT) leader and establish DMT • Identify relevant public/ private sector agencies/ organisations • Establish coordination/ consultative framework and communication systems • Develop, document and communicate disaster management strategy • Education of industry stakeholders, employees, customers and community • Agreement on, and commitment to, activation protocols	*Risk assessment* • Assessment of potential disasters and their probability of occurrence • Development of scenarios on the genesis and impacts of potential disasters • Develop disaster contingency plans

2. Prodromal
When it is apparent that a disaster is imminent
	Mobilisation
	• Warning systems (including general mass media)
	• Establish disaster management command centre
	• Secure facilities
		Disaster contingency plans
		• Identify likely impacts and groups at risk
		• Assess community and visitor capabilities to cope with impacts
		• Articulate the objectives of individual (disaster specific) contingency plans
		• Identify actions necessary to avoid or minimise impacts at each stage
		• Devise strategic priority (action) profiles for each phase
		    ◦ *Prodromal*
		    ◦ *Emergency*
		    ◦ *Intermediate*
		    ◦ *Long-term recovery*
		• On-going review and revision in the light of
		    ◦ *Experience*
		    ◦ *Changes in organisational structures and personnel*
		    ◦ *Changes in the environment*

Table 11.4: Continued.

Phase in disaster process	Elements of the disaster management responses	Principal ingredients of the disaster management strategies
3. *Emergency* The effect of the disaster is felt and action' is necessary to protect people and property	*Action* • Rescue/evacuation procedures • Emergency accommodation and food supplies • Medical/health services • Monitoring and communication systems	
4. *Intermediate* A point where the short-term needs of people have been addressed and the main focus of activity is to restore services and the community to normal	*Recovery* • Damage audit/monitoring system • Clean-up and restoration • Media communication strategy	
5. *Long-term (recovery)* Continuation of previous phase, but items that could not be attended to quickly are attended to at this stage. Post mortem, self-analysis, healing	*Reconstruction and reassessment* • Repair of damaged infrastructure • Rehabilitation of environmentally damaged areas • Counselling victims • Restoration of business/ consumer confidence and development of investment, plans • Debriefing to promote input to revisions of disaster strategies	
6. *Resolution* Routine restored or new improved state establishment	*Review*	

• *Consultation.* To achieve the maximum cohesion, both within the tourism sector and between this sector and the broader community, disaster planning should be based on a consultative process that is both on going and integrated with other areas of strategic planning (e.g. tourism marketing strategies, urban planning and broader regional economic plans). Apart from the bearing plans in these other areas might have on the exposure of the tourism sector to risk and the measures that might be implemented in the response to a disaster, the individuals directly involved change over time and this affects the 'chemistry' of the coordination process.

- *Commitment*. No matter how thoroughly and skilfully the disaster management plan may be developed, and regardless of the level of consultation that takes place in the process, it will be of limited value if the various parties involved are not committed to it and all individuals who are required to take action are not aware of it. As highlighted below, the plan must therefore contain clearly articulated protocols regarding the activation of the strategy and communication/education programme aimed at ensuring that all parties understand what is expected of them.

Ingredients of the tourism disaster management planning process and its outcomes should include:

- *Risk assessment*. An assessment of potential disaster situations that may emerge and their relative probability of occurrence is an essential first step. This should involve an analysis of the history of natural disasters in the region, along with a scanning of the current and emerging environment and alternative scenarios.
- *Prioritisation*. A cascaded strategic priority profile (CSPP), needs to be prepared, with a rank ordering of tasks and activities that need to be undertaken in response to high-risk events identified in the previous step. Part of this process also involves the prioritisation of actions and the articulation of these across organisations so that a coordinated response can be developed. In this context, it needs to be recognised that tourists are vulnerable in unfamiliar surroundings and high priority must be placed on their safety.
- *Protocols*. A clearly enunciated set of protocols to ensure the activities of emergency agencies, tourism authorities and operators are properly coordinated needs to be established and accepted by all parties.
- *Community capabilities audit*. An assessment of the community's capacity to cope with specific types of disasters needs to be carried out so that the appropriate level of emergency relief from external sources can be determined. This should involve an inventory of relevant community (physical, financial and organisational) resources, which is also necessary to address other considerations referred to above.
- *Disaster management command centre*. A properly resourced disaster management command centre, as the focal point for the disaster management team's operations is essential. The location and procedures for setting up this facility must be specified in the plan.
- *Media and monitoring activities*. A media communication strategy involving the early establishment of a centralised source is essential in order to ensure that misleading and contradictory information is not disseminated, and to support the coordination of responses. The media often plays a central role in tourism disaster situations, both in terms of providing important information to tourists during the emergency and in the recovery stage when other sectors of the industry and the consuming public need to be informed about the restoration of services. Systems for monitoring the impacts of disasters, and providing reliable information on safety matters and the status of tourism services are therefore necessary.
- *Warning systems*. Once a disaster strategy is in place, the conditions necessary to activate it must be specified, along with the types of hazard in relation to which it is designed (Huque 1998). Systems for communicating warnings are also important. The

incidence of denial among executives (e.g. 'the floods can't affect us because we are on high ground') highlights the need for definitive warning advice (Drabek 1992).

- *Flexibility.* Certain elements of disaster strategies are applicable to all types of emergencies and might therefore be included as part of a generic framework. However, the exposure of some destinations to certain types of disaster is greater than others and it is essential that these be identified so that responses to the specific impacts and requirements of high-risk events can be planned. Some flexibility is also important, as the precise sequence of actions that are necessary may vary between different types of emergency. Flexibility is also required in the sense that it may be necessary for some organisations to perform functions they do not normally carry out.
- *Involvement, education and review.* The effectiveness of disaster response and recovery plans will be very limited unless those who are required to implement them are directly involved in their development (Quarantelli 1984). Organisations and the community in general need to be informed about the strategy, and the strategy should be periodically reviewed in the light of reactions to it and new developments. Disaster strategies therefore need to be updated and refined on a continuous basis in order to ensure that new information and/or organisational changes are taken into account. In particular, debriefings after disasters have actually occurred are important so that lessons can be learned from experience.

The final point is particularly pertinent with regard to the longer-term disaster preparedness of a destination. It is to be expected that individuals and communities who have experienced a particular type of disaster are better equipped to respond to similar situations in the future, at least to the extent that, with the benefit of hindsight, they have a better knowledge of the actual impacts of the disaster and how to cope with it. However, as Burling & Hyle (1997) have noted, in the case of disaster planning in the United States schools system, few administrators who had actually experienced a disaster transferred the knowledge gained into the disaster strategy development process. This knowledge can only be effectively tapped through a systematic debriefing procedure, with a 'post-mortem' of the event being conducted as a basis for evaluating reactions and refining the strategy.

Conclusion

Natural and human induced disasters alike are neither absolutely predictable nor avoidable. Furthermore, while disasters are, fortunately, relatively rare occurrences and they are to some extent random, it is also true that no destination is immune from such events. In response to the near certainty of experiencing a disaster of some type eventually, tourism organisations can devise means for minimising the damage of, and accelerating the recovering from, such events through the development of disaster management strategies. By studying past events, the responses of those affected and the recovery measures adopted, and retrospectively evaluating the effectiveness of these responses, we can develop strategies for coping with similar events in the future. However, the progress made on this front has been limited because the field has lacked the conceptual framework necessary to structure the cumulative development of knowledge about the impacts of, and effective responses to, tourism disasters. This paper has attempted to address this problem by drawing upon the insights from previous research on disaster (and crisis) management in general, in order to construct a generic model for tourism disaster management specifically.

Within this framework, a distinction has been drawn between crises and disasters. The former have their origins in planning and management deficiencies, and in this sense they are self-inflicted. On the other hand, disasters are triggered by events over which the victim has little control and their impacts are, therefore, to some degree unavoidable. However, the distinction between crises and disasters is often somewhat blurred and it is for this reason it has been suggested that they represent opposite poles of a continuum, rather than a dichotomy. While many disasters are attributable to random natural events, which are beyond the control of the most advanced technology, the impacts of these phenomena can be moderated by planning and management practices. Thus, for instance, various tourism destinations are more or less prone to certain types of natural disasters than others, and in these instances action can be taken to either avoid or at least diminish the harmful effects of the event. Apart from avoiding high-risk locations altogether, one of the more obvious steps that can be taken is to assess the risks an individual destination is exposed to and develop management plans for coping with disaster situations in advance.

A logical step in extending the research described in this paper is to use the framework as a basis for examining and analysing actual cases of tourism disasters. This will enable the generic model to be tested and refined, and provide further insights into the peculiarities of tourism disasters. The methodology for doing this is described in a companion paper, where the case of the Katherine (Australia) flood is examined (Faulkner & Vikulov 2001).

References

Arbel, A., & Bargur, J. (1980). A planning model for crisis management in the tourism industry. *European Journal of Operational Research, 5* (2), 77–85.

Arnold, W. (1980). *Crisis communication.* Iowa: Gorsuch Scarisbrook.

Berke, P. R. (1998). Reducing natural hazard risks through state growth management. *Journal of the American Planning Association, 64* (1), 76–87.

Berman, R., & Roel, G. (1993). Encounter with death and destruction: The 1985 Mexico City earthquake. *Group Analysis, 26,* 89–91.

Blaikie, P., Cannon, T., Davis, I., & Wisner, B. (1994). *At risk: Natural hazards, people's vulnerability and disasters.* London: Routledge.

Booth, S. (1993). *Crisis management strategy: Competition and change in modern enterprises.* New York: Routledge.

Brammer, H. (1990). Floods in Bangladesh: A geographic background to the 1987 and 1988 floods. *Geographical Journal, 156* (1), 12–22.

Burby, R. J., & Wagner, F. (1996). Protecting tourists from death and injury in coastal storms. *Disasters, 20* (1), 49–60.

Burling, W. K., & Hyle, A. (1997). Disaster preparedness planning: Policy and leadership issues. *Disaster prevention and management, 6* (4), 234–244.

Burton, I., Kates, R. W., & White, G. F. (1978). *The environment as hazard*. New York: Oxford University Press.

Cammisa, J. V. (1993). The Miami experience: Natural and manmade disasters, 1992–1993. In: *Expanding responsibilities: A blueprint for the travel industry*. 24th Annual Conference Proceedings of the Travel and Tourism Research Association (pp. 294–295). Whistler, BC.

Capra, F. (1996). *The web of life*. London: Harpers Collins Publishers.

Carter, W. N. (1991). *Disaster management: A disaster manager's handbook*. Manila: Asian Development Bank.

Cassedy, K. (1991). *Crisis management planning in the travel and tourism industry: A study of three destinations and a crisis management planning manual*. San Francisco: PATA.

Chan, N. W. (1995). Flood disaster management in Malaysia: An evaluation of the effectiveness of government resettlement scheme. *Disaster Prevention and Management, 4* (4), 22–29.

Christine, B. (1995). Disaster management: Lessons learned. *Risk Management, 42* (10), 19–34.

Comfort, L. K. (1990). Turning conflict into co-operation: Organizational designs for community response in disasters. *International Journal of Mental Health, 19* (1), 89–108.

Donohue, J. (1982). Some facts and figures on urbanisation in the developing world. *Assignment Children, 57/58*, 21–41.

Drabek, T. E. (1992). Variations in disaster evacuation behaviour: Public responses versus private sector executive decision-making. *Disasters, 16* (2), 105–118.

Drabek, T. E. (1994). Risk perceptions of tourist business managers. *The Environment Professional, 16*, 327–341.

Drabek, T. E. (1995). Disaster responses within the tourism industry. *International Journal of Mass Emergencies and Disasters, 13* (1), 7–23.

Faulkner, B., & Russell, R. (1997). Chaos and complexity in tourism: In search of a new perspective. *Pacific Tourism Review, 1* (2), 91–106.

Faulkner, B., & Vikulov, S. (2001). Katherine, washed out one day, back on track the next: A post-mortem of a tourism disaster. *Tourism Management, 22*, 331–344.

Fink, S. (1986). *Crisis management*. New York: American Association of Management.

Fischer, H. W., & Harr, V. J. (1994). Emergency operating centre response to media blame assignment: A case study of an emergent EOC. *Disaster Prevention and Management, 3* (3), 7–17.

Geipel, R. (1982). *Disaster and reconstruction*. London: Allen and Unwin.

Gleick, J. (1987). *Chaos: Making a new science*. London: Heinemann.

Gonzalez-Herrero, A., & Pratt, C. B. (1998). Marketing crisis in tourism: Communication strategies in the United States and Spain. *Public Relations Review, 24* (1), 83–97.

Granot, H. (1995). Proposed scaling of communal consequences of disaster. *Disaster Prevention and Management, 4* (3), 5–13.

Granot, H. (1997). Emergency inter-organisational relationships. *Disaster prevention and management, 6* (5), 305–310.

Hartmann, B., & Standing, H. (1989). *The poverty of population control: Family planning and health policy in Bangladesh*. London: Bangladesh International Action Group.

Heath, R. (1995). The Kobe Earthquake: Some realities of strategic management of crises and disasters. *Disaster Prevention and Management, 4* (5), 11–24.

Hills, A. E. (1994). Co-ordination and disaster response in the United Kingdom. *Disaster Prevention and Management, 3* (1), 66–71.

Hirose, H. (1982). Volcanic eruption in Northern Japan. *Disasters, 6* (2), 89–91.

Horlick-Jones, T., Fortune, J., & Peters, G. (1991). Measuring disaster trends, Part Two: Statistics and underlying processes. *Disaster Management, 4* (1), 41–44.

Huque, A. S. (1998). Disaster management and the inter-organizational imperative: The Hong Kong disaster plan. *Issues and Studies, 34* (2), 104–123.

Keller, A. Z., & Al-Madhari, A. F. (1996). Risk management and disasters. *Disaster Prevention and Management, 5* (5), 19–22.

Keown-McMullan, C. (1997). Crisis: When does a molehill become a mountain? *Disaster Prevention and Management, 6* (1), 4–10.

Kouzmin, A., Jarman, A. M. G., & Rosenthal, U. (1995). Inter-organisational policy process in disaster management. *Disaster Prevention and Management, 4* (2), 20–37.

Lehrman, C. K. (1986). When fact and fantasy collide: Crisis management in the travel industry. *Public Relations Journal, 42* (4), 25–28.

Lorenz, E. (1993). *The essence of chaos*. Washington: University of Washington Press.

Milo, K. J., & Yoder, S. L. (1991). Recovery from natural disaster: Travel writers and tourist destinations. *Journal of Travel Research, 30* (1), 36–39.

Mitroff, I. (1988). *Break-away thinking*. New York: Wiley.

Murphy, P. E., & Bayley, R. (1989). Tourism and disaster planning. *Geographical Review, 79* (1), 36–46.

Peat, F. D. (1991). *The philosopher's stone: Chaos, synchronicity and the hidden order of the world*. New York: Bantam.

Prigogine, I., & Stengers, I. (1985). *Order out of chaos: Man's new dialogue with nature*. Hammersmith: Flamingo.

Quarantelli, E. L. (1982). Social and organizational problems in a major emergency. *Emergency Planning Digest, 9*, 21.

Quarantelli, E. L. (1984). Organisational behaviour in disasters and implications for disaster planning. *Monographs of the National Emergency Training Center, 1* (2), 1–31.

Quarantelli, E. L. (1996). Local mass media operations in disasters in the USA. *Disaster Prevention and Management, 5* (5), 5–10.

Richardson, B. (1994). Crisis management and the management strategy: Time to 'loop the loop'. *Disaster Prevention and Management, 3* (3), 59–80.

Riley, J., & Meadows, J. (1997). The role of information in disaster planning: A case study approach. *Disaster Prevention and Management, 6* (5), 349–355.

Roberts, V. (1994). Flood management: Bradford paper. *Disaster Prevention and Management, 3* (2), 44–60.

Selbst, P. (1978). The containment and control of organizational crises. In: J. Sutherland (Ed.), *Management Handbook of Public Administrators*. New York: Van Noswand.

Turner, D. (1994). Resources for disaster recovery. *Security Management, ??*, 57–61.

Waldrop, M. (1992). *Complexity: The emerging science and the edge of order and chaos*. Penguin: Simon and Schuster.

Weiner, A. J., & Kahn, H. (1972). Crisis and arms control. In: C. F. Hermann (Ed.), *International Crises: Insights from Behaviour Research* (p. 21). New York: The Free Press.

Young, W. B., & Montgomery, R. J. (1998). Crisis management and its impact on destination marketing: A guide to convention and visitors bureaus. *Journal of Convention and Exhibition Management, 1* (1), 3–18.

Chapter 12

Biting Midges and Tourism in Scotland

Alison Blackwell and Stephen J. Page

Introduction

There is increasing interest in the ways in which tourists interact with the environments they visit as their activities and actions are largely dependent upon their perception of what different natural and man-made environments can provide for leisure. In outdoor environments, the tourist needs to be cognizant of the potential environmental conditions and hazards that may affect the activity space. In many destinations that have a dependence upon the natural environment, such as Scotland, there is a growing trend towards outdoor activities of a passive and active nature as part of the re-imaging by public sector tourism organisations to re-brand the country's appeal and attributes in the competitive strides to improve its market share of inbound tourists. Consequently, this dependence upon the outdoor environment is likely to build an expectation in the visitor's perception of the destination, which needs to be delivered in their experience of the product and environment.

Destinations such as Scotland undoubtedly offer high quality environments for outdoor activities. However, these environments are also associated with potential hazards and risks which tourists may encounter in relation to their health and safety. Such risks exist along a continuum ranging from serious events that affect only a small proportion of visitors (e.g. a fatal accident) to minor events (e.g. insects may be nuisances but potentially affect large numbers of visitors and detract from their enjoyment of a destination and its attractions). A plague of wasps in lowland Scotland in the summer of 2002 became a nuisance and a concern at many outdoor venues to the extent that certain attractions had to issue visitors with warnings and guidelines.

Another widespread and perennial problem in Scotland are *Culicoides spp.* which are biting midges. The biting midge is a small, blood feeding insect that may cause significant distress to both humans and animals through its biting activity. This problem is not confined to Scotland but rather one of global concern. However, there is no better example of the midges' biting attacks on humans than in the Scottish Highlands. Here, one species, *Culicoides impunctatus* dominates and often reaches plague proportions

Managing Tourist Health and Safety in the New Millennium
Copyright © 2003 by Elsevier Science Ltd.
All rights of reproduction in any form reserved.
ISBN: 0-08-044000-2

during the summer months. Although not transmitting disease, the blood feeding habits of the midge can cause significant disruption to outdoor activities, including tourism.

This chapter examines the background to biting midge biology in Scotland, reviewing the nature of its occurrence, prevalence and impact on tourist activity using both anecdotal and documented information regarding the affect of biting midges on visitors to Scotland. Finally, some suggestions for a potential solution to the current difficulties in Scotland are provided.

The Biology and Nature of Midges

Biting midges are known by many different names that vary with geographical location. These include 'sandflies', 'punkies', 'no-see-ums', 'no-nos', 'moose-flies', 'jejens' and 'biting gnats'. Of the inhabited landmasses, the midge is known to be absent from only the southern most areas of South America. Across the globe, there are more than 1,400 named species of the most important genus (Culicoides) highlighting their diversity and occurrence. In New Zealand, midges are a national problem, though particularly significant on the numerous walking tracks on the South Island, such as the famous Milford Track.

The midge's most famous and feared characteristic is its blood feeding which is carried out only by the female who requires the blood components to develop its eggs. The female has specialised mouthparts and uses these to pierce the skin with finely toothed mandibles and maxillae, which work in a scissor-like fashion to create a pool of blood, from which the midge feeds. Saliva, containing an anticoagulant to maintain the flow of blood, is pumped into the wound. The host's body responds to the midge's saliva by an immune reaction of varying degrees, releasing histamine at the site of the wound, resulting in the characteristic itching and swelling at the site of the bite. Left undisturbed, a female midge will feed for three to four minutes and take an average of two millionths of a litre of blood. Differentially, males have less refined mouthparts and feed from only flower nectar and plant juices to obtain an energy source.

Midges in Scotland

The resulting biting nuisance from the midge's blood sucking behaviour can be severe, particularly in Scotland. The Gaelic name for the midge in Scotland is *Meanbh-chuileag* which translates in English to tiny fly. Despite its diminutive size and the fact that the midge is the butt of many a joke, the importance of its impact on outdoor activities, particularly tourism, should not be under estimated. Scotland supports some of the largest populations of biting midges worldwide, particularly in the country's Highland and West Coast areas (Figure 12.1). The density of midges in Scotland represents a significant biting nuisance and problem that is enhanced by their peak activity season of May to September. This period coincides with much of the country's outdoor productive activity including forestry, agriculture and tourism.

Midges are not recent arrivals into Scotland. Indeed they have been found preserved in 75 million-year old amber. It is their more recent history in Scotland, however, that

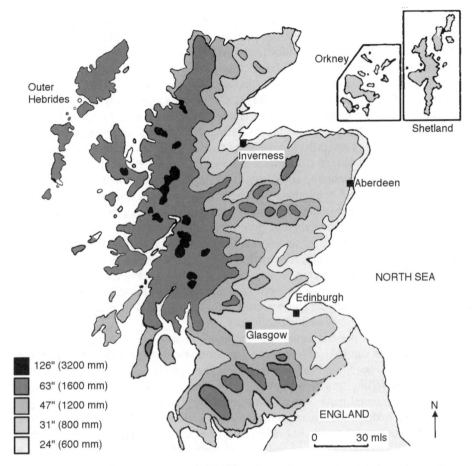

Figure 12.1: The distribution of midges in Scotland.
Redrawn and reproduced with permission.

has been most eloquently described by Hendry (1996), including Bonnie Prince Charlie's encounters with the 'mitches' while hiding in the hills after Culloden. Interestingly, Queen Victoria's unusual relaxation of the rules on smoking at Balmoral was an attempt to deter the midges.

Scotland's midge fauna was first described at the beginning of the twentieth century, but it was not until the 1950s when the Scottish midge problem was more fully investigated with the establishment of the University of Edinburgh's Midge Control Unit. The Unit was responsible for the production of primary knowledge on the biology of midges in Scotland (e.g. Kettle 1951; 1960; 1961). Following a more than 30 year lapse in research, modern approaches to the biting midges in Scotland have recently produced a wealth of new information regarding their ecology and behaviour (see Blackwell 2001).

There appear to be few barriers that *Culicoides spp.* biting midges are unable to surmount. Hendry (1996) reported that three species of Culicoides have been found on St Kilda, beyond the Outer Hebrides in the Atlantic and 100 miles from the Scottish mainland. Of the 37 different species of biting midge in Scotland (Boorman 1986), although a number can co-exist in any one area (Blackwell *et al.* 1992), one species outnumbers all others in most areas. This is *Culicoides impunctatus*, or the 'Highland biting midge' (Figure 12.2). This species is a highly opportunistic feeder and it prefers to take bloodmeals from large mammals (primarily cattle and deer) (Blackwell *et al.* 1994a; 1995), it is not reluctant to take advantage of the rich feeding potential offered by humans. The success of *C. impunctatus* can be related to several features of its population dynamics, including autogeny (the ability to lay a first batch of eggs without taking a bloodmeal) and bivoltinism (two generations each summer) (Blackwell *et al.* 1992). Additionally, environmental factors, such as the wide expanses of the Scottish Highlands which are suitable breeding grounds for this species (Blackwell *et al.* 1994b; 1999) provide the potential for an enormous biomass of this single species of biting midge. During the peak emergence period for *C. impunctatus* as many as 500,000 individuals have been collected, emerging from a two metre square confined area (Blackwell unpublished). Most recently, approximately two kilograms of midges have been caught in single trapping stations each night during the University of Edinburgh's research programme (Blackwell unpublished).

C. impunctatus is present throughout the entire summer in Scotland and its biting attacks on humans can have severe consequences that result in much of the Highland's outdoor industry becoming ineffective during the 'midge season.' These industries include timber felling and tree planting, harvesting, and road works. Forestry and agriculture are integral parts of Scotland's economy. There are many anecdotal reports of the misery midges can cause with these activities but limited numbers of official reports. The Forestry Authority has estimated that of the 65 working days each summer, as much as 20% can be lost due to midge attacks preventing men from working (Hendry & Godwin 1988). The biting attacks on livestock by *C. impunctatus* can have severe consequences with reports of reduced milk yields from cattle and reductions in body mass of red deer (Cameron 1946; Hendry 1996). *Culicoides spp.* are also potential vectors of some lethal pathogens of major livestock industries in Scotland (Jennings & Mellor 1988).

With such strong indications that biting midges have a significant impact on Scotland's economy, it follows that suppression of biting midge activity in Highland and West Coast areas could benefit outdoor industry and recreation significantly improving wealth creation and the quality of life in these areas while increasing the sustainability of rural employment. This is particularly notable with the establishment of Scotland's first national park that straddles the west and central belts of Scotland in the Loch Lomond and Trossachs area where much visitor activity is based outdoors.

Biting midges are not just a problem in Scotland. Significant midge populations act as biting nuisances in areas of England (e.g. the Lakes District, Yorkshire Dales and the Peak District), North Wales and Eire. Elsewhere in the world midges are also biting nuisances in many areas, including the United States of America (e.g. Florida and California), the Caribbean, Mexico, Canada and Australia. In these areas midges can

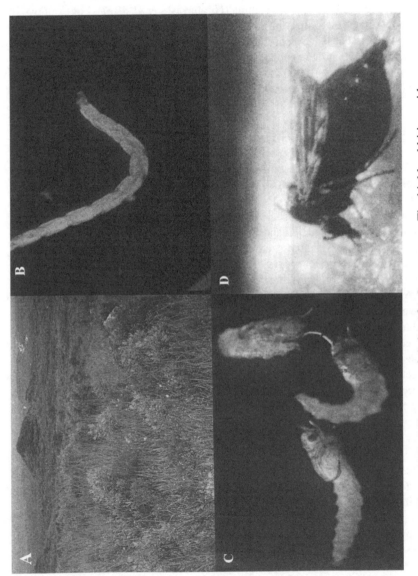

Figure 12.2: Life stages of *Culicoides Impunctatus*: The highland biting midge.

(A) Typical habitat in Western Scotland; (B) Midge larva (found just below the soil surface); (C) Midge pupae (the interface stage between the larva and adult); (D) Blood-fed adult.

also have a significant impact on outdoor recreation and industry alike. For example, in South Carolina it has been estimated that 1,500 work days are lost each year in this way, at an estimated cost of US$250 000 (Haile *et al.* 1984).

Non-biting midges (Chironomidae) can pose problems to tourists in many parts of the world, as a nuisance and may induce asthma (Hirabayashi *et al.* 1997). For example, in a questionnaire survey of tourists in Suwa City near Lake Suwa, Japan to clarify the nuisance for tourists caused by chironomid or non-biting midge, more than 30% of respondents answered that they 'cannot stand any more' massive flights of these insects, demonstrated by more than 10 000 individual adult midges per night being collected by a light trap (Hirabayashi & Okino 1998).

Tourism in Scotland: The Effects of Midge Biting on Tourism

Tourism is one of Scotland's largest industries and employs 177,000 workers and injects £2.47 billion annually into the Scottish economy. In the Highlands and Islands, tourism accounts for 15% of the workforce (The Scotsman 28 November 2000). Ironically, there is far less information available on the effects of midges on the single largest input of income from tourism into the Highlands each summer. It is estimated that 13–14 million tourist trips are made to Scotland each year, with most of these in summer, valued at approximately £2 billion (Hendry 1996). Many caravan parks, campsites and tourist attractions open only during the three months of summer, which dominates the local economy. Although again, most reports of midge attacks are anecdotal, it is clear that midges have the ability to restrict and even prevent many tourist activities. Some visitors are undoubtedly driven away, while others remain and suffer, being mentally unprepared and provided with no immediate relief other than to apply repellents and avoid the peak midge times.

The exact impact of midges in Scotland on 'customer satisfaction' and the loss of income through visitors either leaving, not enjoying their holiday to the full, or persuading friends against a visit to the Highlands are undocumented and the long term consequences for the tourism sector's image are unknown. In fact, the most recent figures suggest that overall in Scotland more than 19 million tourists took overnight trips during 2001 and spent almost £4.1 billion, supporting around 9% of all employment (www.scotexchange.net/KnowYourMarket/kym-essentials.htm). Therefore, any impact of midge activity on tourist spending can have significant economic ramifications, especially in the Highlands and West Coast areas.

Tourism in Scotland is largely, although not wholly, associated with outdoor activities, particularly in those areas most affected by biting midges. These include Argyll, the Islands, Loch Lomond, Stirling/the Trossachs, Ayrshire/Arran and the Highlands. Indeed, according to a 'VisitScotland' (formerly the Scottish Tourist Board) spokesperson around 86% of overseas visitors come to Scotland for the environment (The Scotsman 28 November 2000), supported by an initiative on tourism and the environment (Hughes 1996) and the formation of national park developments, concerned with the integration of conservation and multi-purpose usage. Perhaps the best example is Loch Lomond, the largest area of fresh water in Great Britain, having diverse communities of plants and animals together with a high amenity value to tourists for recreational purposes, as a fishery and a source of water supply. These characteristics have led to the suggestion that Loch Lomond will be the 'Jewel in the Crown' of the proposed Loch Lomond and Trossachs National Park (Maitland *et al.* 2000).

Visitors to Scotland are involved in a range of activities, including:

- Cycling (£73m expenditure);
- Walking (accounting for 20% of Scotland's tourist expenditure; at least £438m);
- Angling (with a £13m expenditure in the Borders areas of Scotland alone); and
- Golfing (worth £100m to the Scottish economy, with golfers spending more per trip than general holiday makers; £68/day (United Kingdom) — £94/day (French visitors).

Visitors taking part in these activities come from a variety of destinations and utilise a range of accommodation, including camping and caravans (34% of visitors, spending £6 per day on accommodation) and self-catering (31% visitors; £12–16/day). At the upper end of the scale are those staying in hotels (£31-£38/day)

Many of these activities represent niche markets. These specific segments of the market have well-defined tourism products and marketing tailored to meet the interests of the visitor. According to 'VisitScotland', customers increasingly are looking for a range of experiences and it is recommended that businesses need to work together to offer a variety of activities. For example, a visitor on a walking holiday may wish to golf and visit gardens (VisitScotland Tourism Futures: www.scotexchange.net/knowyour-market/nicheops.htm). Of those niches being developed by VisitScotland, the majority involve outdoor activities that, during the summer months, could involve interactions with biting midges in some areas (Table 12.1).

Cyclists holidaying in Scotland are an ideal group to consider in relation to possible interactions with midges, in the areas visited, chosen accommodation and influences determining their choice of holiday. The region of origin profile of visitors who go cycling/mountain biking is similar to that of all holidaymakers, with visitors from England being the most prevalent, followed by those from Scotland and, in turn, Europe. Two adult groups make up a large proportion of the holiday sector that go cycling, but families are significant in the Highlands and Argyll and Bute, as are single adults in the Western Isles. Visitors who go cycling on holiday are more likely than holidaymakers in general to have visited the area previously. A previous visit and/or referral from friends and relatives are a key influence, while tourist brochures and guidebooks are important secondary influences (www.scotexchange.net).

Cyclists undoubtedly do suffer from Scottish midges during the summer months, with a number of websites giving cycling information including appropriate warnings. These include: "An added danger to the cycling in Scotland is the midges. This pest is a gruesome, bloodletting reality and not a joke" (Cycling in Scotland www.infernomusic.co.uk/cycling.html) and "This is Scotland and at times there is no avoiding the fabled midge . . . try to get your days walking over with by late afternoon. The midge enjoys dining in the early evening!" (Walking the Southern Upland Way www.Southernuplandway.com). Since the internet is becoming a powerful advertising

Table 12.1: Out-of-doors niche tourist activities in Scotland and prospects for their development.

Niche Activity	Annual Expenditure (£m)	Prospects
Golf	100	Excellent
Walking	438	Excellent
Cycling	73	Very good
Wildlife	57	Very good
Field sports	53	Good
Cruising	12	Good
Sailing	10	Good
Equestrian	Nil	Good

(VisitScotland Tourism Futures
http://www.scotexchange.net/knowyourmarket/nicheops.htm)

and marketing tool for tourism organisations and activities in Scotland (Cano & Prentice 1998a; 1998b), such reports are potentially of significance. Indeed, the Highlands and Islands of Scotland, where the highest numbers of midge numbers are found, is the tourist region with the highest number of World Wide Web sites (145 in 1998) (Cano & Prentice 1998a).

According to 'VisitScotland Tourism Futures', Scotland has the potential to underpin a prosperous and a fast-growing golf tourism industry and for this reason, the Scottish Golf Tourism Executive, comprising the main golf and tourism agencies and the private sector, has been established to develop this national strategy and lead its implementation. An action plan covering the priority areas of development, service delivery, promotion and research and monitoring has been agreed, and the key measures to be taken set out, with the aim of providing a guide to developing and packaging Scottish golf. It is understood that the approach must be consumer-led, providing the quality of information that customers need, available where and when they want it.

Recommendation by friends and family, together with previous experience, are the two primary influences in the visitors' choice of golfing destination "ensuring that the quality of the Scottish golf experience lives up to customers' highest expectations is essential if we are to win positive referrals and repeat visits" (www.scotexchange.net/KnowYourMarket/Niche/golf2.asp). Although this will clearly concentrate on the provision of appropriate tee times for visiting golfers, improving accessibility to famous and lesser known courses, a substantial proportion of the more than the 500 courses in Scotland are found in the heart of 'midge territory'. Some internet sites do emphasize the claim that their courses are totally 'midge free' and others state that the appropriate anti-midge precautions have been adopted.

Tourism in Scotland and Climate

According to 'The Scotsman' newspaper, the most common feature of visitor complaints in Scotland is the weather. The article suggested that although tourists complain about the poor climate, they do not come to Scotland for beach holidays and dazzling blue skies, but are drawn by the scenery and vast expanses of untouched countryside (The Scotsman 28 November 2000). Since much of Scotland's tourism relies on the environment, what effects might predict changes in the country's climate through global warming have on the associated activities? In Scotland, the indications are that winters are becoming milder and summers drier. This may threaten the financial viability of tourism related enterprises during the winter months, primarily skiing, potentially having a profound impact on towns such as Aviemore, which depends almost wholly on skiers visiting the nearby Cairngorms for much of its winter income. Wetter spring and autumn periods may threaten other outdoor activities, such as angling. With peak fishing periods of February to March and October to November, times of the year which could be adversely affected by heavy rainfall and flooding in the river, there could be significant deterioration of fish stocks, removing several million pounds of tourism income. It is possible that Scottish tourism may reap some localised benefits from these ongoing climate changes, particularly with warmer summers and the predicted reduction in dull and damp 'dreich' summer days in Upland areas (Harrison *et al.* 1999). The midge, however, might benefit from climate change. Although unlikely to see the increase in livestock diseases that other European countries might experience (e.g. recent incursions of 'bluetongue virus' of sheep into Sardinia and Southern Italy from North Africa), the midge 'season' could become an even greater problem to tourists as temperatures and rainfall rise. This could result in larger areas of potential breeding grounds, midges living for longer during the summer time and perhaps even a shift from Highland and West Coast areas of Scotland into more easterly areas. Indeed, with a warm, wet summer during 2002, there have been a number of reports of midges causing significant problems in the centre of Edinburgh on the East coast of Scotland, not normally affected by midges.

Tourist Health and Scotland: Alerting Visitors to the Midge Issue

Visitors travelling abroad often expect to receive information on potential health hazards in their destination. There is no better example than visitors travelling to countries where they are likely to come into contact with disease transmitting-mosquitoes. Health advice can range from using personal repellents and taking anti-malarial tablets prior to, and during visits to potential malaria zones (Schopke *et al.* 1998) to vaccinate against mosquito-transmitted viruses.

While biting midges are confirmed vectors of some of the most lethal diseases of livestock (although not in the United Kingdom), pathogen transmission to humans by biting midges, however, appears to be minimal. Exceptions include transmission of several species of filarial worms in tropical and sub-tropical regions (e.g. Mansonella ozzardie, transmitted by *C. phlebotomus* in coastal North Trinidad (Nathan 1981) and a small number of viruses, including Oropouche, which is strongly pathogenic to humans in parts of South America and the West Indies (Linley *et al.* 1983). The bites of midges alone, however, can pose a potential health risk. When bitten, humans' skin reaction is usually mild, and symptoms include temporary burning and slight swelling. More sensitive individuals may develop weals, blisters and extreme tenderness of the skin. In very few cases, extreme allergies can lead to hospitalisation. The scratching of bites can lead to scarring and secondary infection.

The midge bite in Scotland is so infamous that it has become the subject of a dedicated section on tourism health in Scotland in a popular internet site covering a range of tourism activities in the Highlands: "an aggressive breed of midge (very similar to a gnat) is very common in most parts of Scotland, more so in wooded areas by fresh water (areas common in Scotland). They breed between April and October, and at their worst at the start and end of the day. A bite from a midge is pretty much harmless but it will leave you with a hell of an itch for a while! To ensure that you suffer only the minimum of bites, apply insect repellent such as antihistamine cream. If they really are a nuisance you may want to invest in a midge net" (www.highlandtraveller.com/travel/health.htm). Such portrayals of the midge indicate that the tourism industry needs to be more positive in its recognition, identification and warnings to visitors so that the health implications of such endemic problems are minimised. There are undoubtedly impacts that arise from such activity.

The Impact of Biting Midge Activity on Tourism in Scotland

Scotland supports some of the largest populations of biting midges worldwide. Although not transmitting disease in Scotland, the pure density of midges represents a significant problem, which is compounded by midge activity and tourist season of May to September.

The Moffat Centre's Visitor 'Attractions Barometer' based at Glasgow Caledonian University, estimated that visits to 229 Scottish attractions declined 8% in August 2000 compared to 1999 and more recently, recorded a 2.6% decline in visitors for 2001 with the Highlands of Scotland suffering the second largest decline of 21.7%. Although there are clearly many contributory factors to Scotland's declining tourist industry, anecdotal reports and a recent survey of tourists in Scotland indicate that biting midge attacks can curtail the number and duration of tourist visits. Recent research indicates that tourists would welcome additional information about the midge and take appropriate precautions. Of those first time visitors to Scotland, only 41% were aware of the existence of biting midges prior to their visit; 62% would be deterred from visiting at the same time of the year again and 86% would warn off their friends and family (Blackwell 2000).

There are numerous examples of the disruption caused by biting midge activities to major tourist events in Scotland. In September 2001, an outdoor performance of Shakespeare's 'The Tempest' was cancelled due to increasing midge problems. The performing group was apparently 'bitten to death' in some scenes during a number of outdoor performances before withdrawing from a performance on Rannoch Moor, close to one of the epicentres of the Scottish midge population in Glencoe. The reason given was that the actors would find it difficult to give convincing performances under these circumstances and that it would be feasible to admit defeat.

Numerous individuals have recounted their own horror stories of their encounters with midges during their holiday in Scotland. Extracts of stories submitted to a website investigating the attacking behaviour of the Highland biting midge (www.midgepro-ductions.com) include:

> On holiday in Glen Affric . . . wondered why people would not get out of their cars . . . On stepping out of the car was surrounded by huge clouds of midges which followed us to over three thousand feet and back to the car' and 'My sleeping bag was no longer yellow in colour but black due to all the midges that were on it . . . It has been the longest night of my life and I will certainly never go camping in the summer again . . .

Then there was the family who were trapped inside their holiday cottage for two whole days unable to escape; the people who ran from a campsite pursued by a cloud of midges and did not stop for two hours; a French couple who had no warning at all of the highland midge, and spent the whole night with a cigarette lighter killing their tent invaders; and another French woman who was perhaps given the best piece of advice before leaving home: 'if you kill one midge, a hundred come to the funeral.' Some of these stories are associated with some extremely graphic images (Figure 12.3) and a number of warnings to potential visitors, from both the United Kingdom and overseas. The majority of these are included in the ever-increasing number of Internet sites about Scotland including:

- www.scotland.com/culture/facts/climate.html "For the rest of the summer months, these tiny, biting insects can cause some discomfort if you are by water, in wooded areas or on the hills".
- www.islesbritannic.com "The Highland Midge is the most voracious of the Scottish midges . . . What makes them particularly annoying is that they do not attack in ones — they attack in force. Once a female finds a victim, she sends out a pheromone which signals to other midges. The Highland Midge is most prevalent in areas that receive more than 50 inches of rain per year. In Scotland that means Argyll and the rest of the West tend to have the healthiest Highland Midge populations".

Interestingly, in consulting the 'VisitScotland' website and those of the separate area Tourist Boards, there are no obvious mentions of the midge, potentially explained by the delicate nexus between public relations and tourism, where negative publicity is said to damage the image of the destination and may deter visitors from even planning a holiday. This argument has been widely cited by interest groups within the tourism industry, particularly given the effects of large scale crises such as the Foot and Mouth epidemic in the United Kingdom on overseas visitors and the aftermath of the 11 September 2001 attacks. Only visitors who actively research their travel plans fully via the World Wide Web are likely to receive alerts to the potential midge problem.

In an attempt to provide some preliminary quantification on the disruption caused to tourists, recreationalists and residents at leisure in Scotland by biting midges, Blackwell

Figure 12.3: Midge man interactions in Scotland.

(A) Midge Bait (www.midgeproductions.com); (B) A 'handful' of Scottish biting midges; (C) One-night's midge catch (approximately 2 kg), University of Edinburgh Midge Research Programme 2002; (D) Midge protection.

(2000) undertook a questionnaire survey to scope some of the principal issues and effects of midge biting on the Scottish visitor population. In addition to examining the nature of visitor's views on the midge, they were asked about what measures they took to alleviate any suffering caused by midge biting attacks. A total of 2,000 questionnaires were distributed in the Highland and West Coast areas of Scotland, concentrating on caravan and campsites, youth hostels and other tourist areas. The islands of Skye and Islay were included. The questionnaires were circulated during July and August 1999, at a time that coincided with school holidays and the peak periods for tourists and midges. Although only a very limited survey and biased towards people involved with outdoor activities, the study provided some interesting data.

Among the respondents (35% of those asked), the majority (80%) had been bitten by midges within the last two days, reporting an average reaction (such as swelling, discomfort and itching) of 5.6 ± 0.3 (scored out of 10, with 10 being the greatest reaction and 1 being the least). This type of response appears to closely mirror the international studies of tourist health, which indicate that insect bites are a common problem experienced by visitors (Page 2002).

Despite this relatively high response, only 57.6% stated that they used repellents, perhaps related to the fact that 75% of respondents thought that repellents were either 'fair', 'not very good' or 'useless' at protecting them from midge bites. The use of insect repellents and a number of other 'midge avoidance' tactics were stated, encompassing both the common-sense actions (e.g. avoidance of peak midge periods, remaining inside and wearing protective clothing) and the bizarre, such as eating copious amounts of garlic. When respondents classifying themselves as 'local' to the study area were asked for an assessment of how much midges affected their lifestyle and activities during the summer months, the mean response was 5.2 ± 5.7 (scored out of 10, with 1 being the least affect and 10 being the greatest). Activities stated as being most affected included all outdoor activities, with agriculture being particularly highlighted in a number of recreational activities including hill walking, gardening, horse riding and camping. Of the tourists who completed questionnaires, 22% were from Scotland, 34% from England, 3% from Wales, 1.5% from the United States of America and 39.5% from other countries. Some 31% of the tourists camped, 38% stayed in hotels, bed and breakfast or self-catered accommodation, with the remainder in other forms of accommodation.

Seventy percent of respondents had visited Scotland one or more times previously and approximately the same proportion had been aware of midges in Scotland before their holiday. Of those visitors who had not previously visited Scotland, only 41% were aware of the existence of biting midges prior to their visit. Data concerning whether or not the midges would deter visitors re-visiting at the same time of the year again, or whether they would think of telling their friends and family to avoid areas and times of Scotland where midges may be a problem was extremely interesting (Table 12.2). Nearly half of those responding stated that the midges would put them off from returning to Scotland at the same time of the year again and 68% stated that they would think about warning their friends and family to avoid the midge season. Furthermore, taking only those tourists who stated that they had not visited Scotland previously, these figures increased to 62% for those who would be deterred from visiting at the same time

Table 12.2: The reaction of tourists to biting midge activity (Blackwell 2000).

Aware of midges before the visit?		Put off by midges from returning to Scotland at the same time of the year again?		Warn friends and family to avoid the midge season in Scotland?		Overall effect of midge activity on the visit
Yes	**No**	**Yes**	**No**	**Yes**	**No**	**Mean Score/10**
69.1%	30.9%	48.5%	51.5%	67.6%	32.4%	5.0 ± 3.4

of the year again and 86% who would warn off their friends and family. Overall, the effects of biting midges on various tourist activities was scored with a mean of 5.0 (scored out of 10), rising to 5.8 for those who had not previously visited Scotland. A range of outdoor sports and recreational activities were highlighted as most affected, including hill walking, climbing, camping, playing golf and generally being outside in the early morning and evening. A number of specific comments were recorded by the tourist respondents, with many reflecting the feelings behind this particular comment: "I visit Scotland many times in the year but not normally in summer — July, August — I will not come again then" (Blackwell 2000).

Perhaps the above comments are best summarised in one statement found on a BBC website: "One of the main exports of the area is called 'Tourists with Midge Bites'. Midge bites are not a new form of vegetable spread for toast, but rather little red spots congregating in the nether regions. The Midge (pronounce midgee) is a small fly that hangs around the lochs of the area, feeding off unwary tourists and fishermen" (www.bbc.co.uk/dna/h2g2/A198614). Such a statement embodies the perennial problem which visitors and locals confront during summer.

The Way Forward?

Clearly those actively involved with biting midge research in Scotland and interacting with tourists who have suffered from midge attacks recognise that it would be beneficial to raise the profile of the Scottish biting midges so that first time visitors are aware of the possible problems they face and can be better prepared. The use of insect repellents is often not satisfactory but their use and effectiveness, however, could be optimised by informing visitors when and where they are most likely to encounter midges and what measures they can take to avoid them. The tourist industry has traditionally omitted or down played the presence of midges in Scotland. It is suggested, however, that this is changing, with scanning electron micrographs of *C. impunctatus* appearing in the Highland Tourist Board's 2000 brochure (The Scotsman 13 May 2000). Some specialist walking and climbing clubs are beginning to publicise information on midge biology and strategies for protection from biting (e.g. www.gpsinternet.com/NMC/midges.bio.html).

Midges have recently become an art subject of significant public interest. Resulting photographic and video work can be seen at www.midgeproductions.com. Rather than deter visitors from Scotland, it might be anticipated that such websites will assist people to make informed decisions regarding areas to visit and appropriate anti-midge precautions. The result could be improved customer satisfaction, associated with benefits to the local economies, including increased visitor spending as a result of:

- Visitors being less confined (increased knowledge of midges) to places they can visit/times when they are able to do so; and
- Fewer environmental constraints to ecotourism and environmental tourism activity (both growth sectors of the Scottish tourism sector, including walking and cycling). Job creation could follow from this with current data on Scotland and ecotourism suggesting employment of one person per £20,000 of visitor spending in remoter areas (see Page & Dowling 2001).

More proactively, can midge numbers be reduced to a level, which, on a localised scale, would significantly reduce the biting pressure on humans? Clearly, what is required is the construction of cost-benefit analysis models of the suppression of midge activity in key areas, quantifying the potential economic benefits from reducing midge numbers by certain proportions at key sites. The minimum level of midge suppression which would significantly improve wealth creation and the quality of life in these areas could eventually be fed back into the design of midge management programmes, for which the groundwork has largely been carried out during the past decade of research on *C. impunctatus* in Scotland (Blackwell 2001).

Midge management programmes are currently the subject of further research and development activity. To date, there has not been an effective method of midge control or one that is environmentally sound. Early attempts at insecticidal fogging against the adult midge stages met with little success (Kettle 1949), probably because the insects avoided contact with the insecticide by hiding beneath vegetation. Furthermore, the broadcast spraying of insecticides could lead to the rapid development of resistant strains, along with significant environmental and health risks. These factors make the broadcast spraying of insecticides against larval midges unacceptable, despite the recent development of insecticides that are less persistent and broad-spectrum than the organochlorines, such as DDT, which had larvicidal effects against *C. impunctatus* in earlier trials (Kettle *et al.* 1956; Kettle & Parish 1957). There is, however, a growing trend towards the use of botanical products, such as neem-based products (Mulla & Su 1999), for the control of insects of medical and veterinary importance and a number of natural products have the potential for midge larval control (Blackwell unpublished). Concerning habitat manipulation to make areas less appealing to midges for breeding, Hendry (1996) suggests that it is doubtful that any attempt to alter the landscape of localised areas would, in the long term, be successful against midges in Scotland. Considering the relatively high mobility of these insects and the vast areas of potential breeding grounds, this is probably true.

One way forward might be the development of new, more effective repellents. A range of chemical repellents, in various formulations sold under numerous trade names are currently the main defence against midges in Scotland. The most widely used chemical in insect repellents is DEET (*N,N*-diethyl-*m*-toluamide), forming the main active constituent in the majority of 'over-the-counter' preparations, with concentrations varying from 10% to 90%. Since it was first marketed in 1956, DEET has remained the most effective repellent against midges, mosquitoes and other biting pests. These products are effective if applied regularly, but how safe are they? Toxicity from casual use is thought to be low and there are no definite reports that these products are not safe if used correctly and sensibly. There are, however, a number of reports of toxic effects of DEET with long-term use. The nervous system, immune system and skin are the prime targets for adverse reactions, with dermal absorption of DEET reported to be three to 8% for humans (Selim *et al.* 1995). A total of 9,086 human exposures to DEET were reported to the U.S. Poison Control Centers from 1985 to 1989 (Veltri *et al.* 1994) and DEET exposure has been cited as partly responsible for psychological effects associated with 'Gulf War syndrome' (Haley & Kurt 1997). Children may be particularly vulnerable, experiencing seizures (Lipscomb *et al.* 1992) and dermatitis (Wantke *et al.* 1996), thus prompting the search for safer alternatives (Mafong & Kaplan 1997). There is a long history of the use of herbal concoctions against midges, including oils distilled from a range of plants, including lemon grass, eucalyptus, cypress, lavender and thyme. A common factor for most of these is that they contain a number of terpenoid compounds, such as citronella and limonene. Recently, a number of essential oils and other plant-derived preparations have been investigated more fully for insect repellent activity, including eucalyptus-based repellents against *C. impunctatus* (Trigg 1996), oregano oil against the biting midge *Culicoides imicola* (Braverman & Chizov-Ginzburg 1997) and salicyclic acid and its derivatives against *C. impunctatus* (Stuart *et al.* 1999).The oil derived from the leaves of Myrica gale L. (bog myrtle), a deciduous shrub that grows widely in the Highlands of Scotland has been evaluated as an insect repellent against *C. impunctatus*, revealing repellent activity equal to that of DEET (Stuart 1990; Evans *et al.* 1996; Stuart & Stuart 1999).

Finally, recent research into the ways in which midges find their bloodmeal hosts has led to the development of a number of effective trapping systems, effectively mimicking a living, breathing mammal, drawing midges towards the trap with a combination of carbon dioxide (the main attractant for a blood-seeking midge) and other host-related attractants. Recent trials of such traps conducted by the University of Edinburgh have trapped approximately three million midges per night (Mands 2002). Figure 13.2C provides an example of these. It is probable that in the not too distant future that every visitor to Scotland may see such a trap sitting in the garden of their hotel or campsite.

Recommendations

In strategic terms, there is clearly a need for a more proactive approach by the public sector agencies responsible for the marketing, promoting and management of tourism in Scotland to recognise the positive economic benefits of a midge management programme for those areas that are heavily dependent upon tourism and recreational

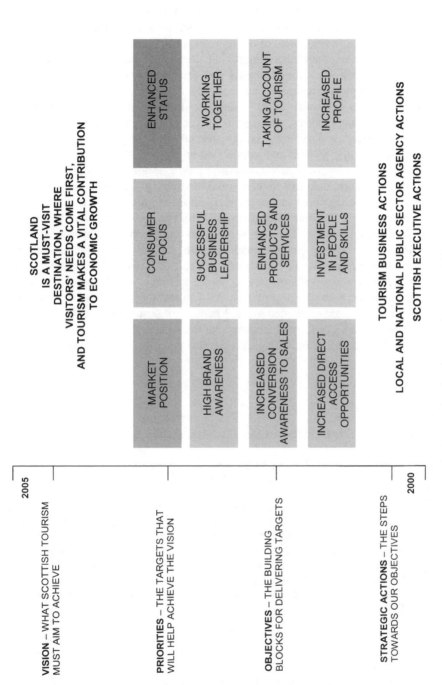

Figure 12.4: Tourism framework for action 2002–2005.

activity for their economic mainstay. In the recent Tourism Framework for Action 2002–2005 (Scottish Executive 2002), a new vision and set of priorities have been established to guide the development of the tourism sector (see Figure 12.4).

These can best be summarised as improving the market position, with increased consumer focus and clearly defined brands and products. Above all, the aim is to enhance the status of the tourism sector and a series of actions are listed in Figure 12.4. Implementation of the Framework for Action is a shared responsibility between the public and private sector agencies, and there is a clear role for improving the quality of the visitor experience by developing midge management programmes. This would enhance the nature of the product and satisfaction levels, and increase word-of-mouth recommendations given by the highly competitive nature of attracting overseas visitors. If visitors are to be encouraged to visit rural districts which have strong midge infestations, better management strategies and unlocking the potential of the huge knowledge base that exists within the university sector can directly assist in knowledge transfer and problem resolution. This may seem all too simple, but the investment in the commercialisation of existing technological capability to assist in midge forecasting, modelling and management would directly add value, address a constant bugbear in the tourism sector and harness a technological expertise that may be able to be exported to other parts of the world with similar problems. As with research discussed in many other parts of this book, there are direct commercial benefits of sharing and commercialising research on tourism that can directly improve the tourist's well-being and health, even where the tourism industry is unwilling to recognise the negativity of admitting that a problem exists.

References

Blackwell, A. (2000). Scottish biting midges: Tourist attraction or deterrent? *Antenna*, *24*, 144–150.

Blackwell, A. (2001). The Scottish biting midge, *Culicoides Impunctatus Goetghebuer*: Current research status and prospects for future control. *Veterinary Bulletin* (November 2001), *71*, 1R–8R.

Blackwell, A., Mordue (Luntz), A. J., Young, M. R., & Mordue, W. (1992). Bivoltinism, survival rates and reproductive characteristics of the Scottish biting midge, *Culicoides Impunctatus* (Diptera: Ceratopogonidae) in Scotland. *Bulletin of Entomolical Research*, *82*, 299–306.

Blackwell, A., Mordue (Luntz), A. J., & Mordue, W. (1994a). Identification of bloodmeals of the Scottish biting midge, *Culicoides Impunctatus*, by indirect enzyme linked immunosorbent assay (ELISA). *Medical and Veterinary Entomology*, *8*, 20–24.

Blackwell, A., Young, M. R., & Mordue, W. (1994b). The microhabitat of *Culicoides Impunctatus* (Diptera: Ceratopogonidae) larvae in Scotland. *Bulletin of Entomological Research*, *84*, 295–301.

Blackwell, A., Brown, M., & Mordue, W. (1995). The use of an enhanced ELISA method for the identification of *Culicoides Bloodmeals* in host-preference studies. *Medical and Veterinary Entomology*, *9*, 14–218.

Blackwell, A., Boag, B., Gordon, S. C., Lock, K. A., & Marshall, B. (1999). The spatial distribution of *Culicoides Impunctatus* biting midge larvae. *Medical and Veterinary Entomology*, *13*, 362–371.

Boorman, J. (1986). British *Culicoides* (Diptera: Ceratopogonidae): Notes on distribution and biology. *Entomological Gazette, 37*, 253–266.

Braverman, Y., & Chizov-Ginzburg, A. (1997). Repellency of synthetic and plant-derived preparations for *Culicoides Imicola*. *Medical and Veterinary Entomology, 11*, 355–360.

Cameron, A. E. (1946). Report on midge repellents: In: *Control of midges: An interim report of a sub-committee of the scientific advisory committee, Department of Health, Scotland* (pp. 4–8). Edinburgh: HMSO.

Cano, V., & Prentice, R. (1998a). WWW Homepages for the tourism industry: The Scottish experience. *Aslib Proceedings, 50*, 61–68.

Cano, V., & Prentice, R. (1998b). Opportunities for endearment to place through electronic 'visiting': WWW Homepages and the tourism promotion of Scotland. *Tourism Management, 19*, 67–73.

Evans, K. A., Blackwell, A., Deans, S. J., & Simpson, M. J. (1996). The repellent properties of myrica gale to haematophagous insect pests of man. In: K B. Wildey (Ed.), *Proceedings of the 2nd International Conference on insect pests in the urban environment* (pp. 427–430). Exeter: BPC Digital Techset Ltd.

Haile, D. G., Kline, D. L., Reinert, J. F., & Biery, T. L. (1984). Effects of aerial applications of naled on *Culicoides* biting midges, mosquitoes and tabanids on Parris Island, South Carolina. *Mosquito News, 44*, 178–183.

Harrison, S. J., Winterbottom, S. J., & Sheppard, C. (1999). The potential effects of climate change on the Scottish Tourist industry. *Tourism Management, 20*, 203–211.

Haley, R. W., & Kurt, T. L. (1997). Self-reported exposure to neurotoxic chemical combinations in the Gulf War — a cross-sectional epidemiologic study. *Journal of the American Medical Association, 277*, 231–237.

Hendry, G. (1996). *Midges in Scotland*. Edinburgh: Mercat Press.

Hendry, G., & Godwin, G. (1988). Biting midges in Scottish forestry: A costly irritant or a trivial nuisance? *Scottish Forestry, 42*, 113–119.

Hirabayashi, K., & Okino, T. (1998). Massive flights of Chironomid midge nuisance insects around a hypereutrophic lake in Japan: A questionnaire survey of tourists. *Journal of the Kansas Entomological Society, 71*, 439–446.

Hirabayashi, K., Kubo, K., Yamaguchi, S., Fujimoto, K., Murakami, G., & Nasu, Y. (1997). Studies of bronchial asthma induced by Chironomid Midges (Diptera) around a hypereutrophic lake in Japan. *Allergy, 52*, 188–195.

Hughes, G. (1996). Tourism and the environment; A sustainable partnership. *Scottish Geographical Magazine, 112*, 107–113.

Jennings, D. M., & Mellor, P. S. (1988). The vector potential of British Culicoides species for bluetongue virus. *Veterinary Microbiology, 17*, 1–10.

Kettle, D. S. (1949). An attempt to control *Culicoides Impunctatus Goetghebuer* in Scotland by barrier-spraying. *Annals of Tropical Medicine and Parasitology, 43*, 284–296.

Kettle, D. S. (1951). The spatial distribution of *Culicoides Impunctatus Goet*. Under woodland and moorland conditions and its flight range through woodland. *Bulletin of Entomological Research, 42*, 239–291.

Kettle, D. S. (1960). The flight of *Culicoides Impunctatus Goetghebuer* (Diptera, Ceratopogonidae) over moorland and its bearing on midge control. *Bulletin of Entomological Research, 51*, 461–490.

Kettle, D. S. (1961). A study of the association between moorland vegetation and breeding sites of *Culicoides* (Diptera, Ceratopogonidae). *Bulletin of Entomological Research, 52*, 381–411.

Kettle, D. S., Nash, R. G., & Hopkins, B. A. (1956). Field tests with larvicides against *Culicoides Impunctatus Goetgh*. in Scotland. *Bulletin of Entomological Research, 47*, 533–573.

Kettle, D. S., & Parish, R. H. (1957). Field trials of larvicides against culicoides with a discussion on the relationship between rainfall and larval control. *Bulletin of Entomological Research, 48,* 425–434.

Linley, J. R., Hoch, A. L., & Pinheiro, F. P. (1983). Biting midges (Diptera: Ceratopogonidae) and human health. *Journal of Medical Entomology, 20,* 347–364.

Lipscomb, J. W., Kramer, J. E., & Leiken, J. B. (1992). Seizure following brief exposure to the insect repellent DEET. *Annals of Emergency Medicine, 21,* 315–317.

Mafong, E. A., & Kaplan, L. A. (1997). Insect repellents: What really works? *Postgraduate Medicine, 102,* 63.

Maitland, P. S., Adams, C. E., & Mitchell, J. (2000). The natural heritage of Loch Lomond: Its importance in a national and international context. *Scottish Geographical Journal, 116,* 181–196.

Mands, V. (2002). *The application of kairomones for the control and monitoring Culicoides spp. in Scotland.* Unpublished masters thesis, University of Edinburgh, Scotland.

Mulla, M. S., & Su, T. Y. (1999). Activity and biological effects of neem products against arthropods of medical and veterinary importance. *Journal of the American Mosquito Control Association, 15,* 133–152.

Nathan, M. B. (1981). Transmission of the human filarial parasite Mansonella Ozzardi by *Culicoides Phlebotomus* (Williston) (Diptera, Ceratopogonidae) in coastal North Trinidad. *Bulletin of Entomological Research, 71,* 97–105.

Page, S. J. (2002). Tourist health and safety. *Travel and Tourism Analyst, 4,* 1–36.

Page, S. J., & Dowling, R. (2001). *Ecotourism.* Harlow: Pearson Education.

Schoepke, A., Steffen, R., & Gratz, N. (1998). Effectiveness of personal protection measures against mosquito bites for Malaria Prophylaxis in travellers. *Journal of Travel Medicine, 5,* 188–192.

Scottish Executive (2002). *Tourism framework for action 2002–2005.* Scottish Executive: Edinburgh

Selim, S., Hartnagel, R. E., Osimitz, T. G., Gabriel, K. L., & Schoenig, G. P. (1995). Absorption, metabolism, and excretion of DEET following dermal application to human volunteers. *Fundamental and Applied Toxicology, 25,* 95–100.

Stuart, A. E. (1990). Paralysis of C. Impunctatus after exposure to oil of Myrica Gale. *Proceedings of the Royal College of Physicians of Edinburgh, 20,* 463–466.

Stuart, A. E., & Stuart, C. L. E. (1999). A microscope slide test for the evaluation of insect repellents as used with Culicoides Impunctatus. *Entomologia Experimentalis et Applicata, 89,* 277–280.

Stuart, A. E., Brooks, C. J. W., Prescott, R. J., & Blackwell, A. (1999). The repellent and antifeedant activity of salicylic acid and related compounds against the biting midge, *Culicoides Impunctatus* (Diptera: Ceratopogonidae). *Journal of Medical Entomology, 32,* 222–227.

Trigg, J. K. (1996). Evaluation of a eucalyptus-based repellent against *Culicoides Impunctatus* (Diptera, Ceratopogonidae) in Scotland. *Journal of the American Mosquito Control Association, 12,* 329–330.

Veltri, J. C., Osimitz, T. G., Bradford, D. C., & Page, B. C. (1994). Retrospective analysis of calls to posion control centers resulting from exposure to the insect repellent DEET from 1985–1989. *Journal of Toxicology -Clinical Toxicology, 32,* 1–16.

Wantke, F., Focke, M., Hemmer, Gotz, M., & Jarisch, R. (1996). Gerneralised Urticaria induced by a DEET-containing insect repellent in a child. *Contact Dermatitis, 35,* 186–187.

Chapter 13

Tourist Safety and the Urban Environment

Michael Barker, Stephen J. Page and Denny Meyer

Introduction

Urban tourism, which is broadly defined as travel to towns and cities, provides the context for a diverse range of social, cultural and economic activities in which the population engages, and where tourism leisure and entertainment form major service activities (Page 1995; Page & Hall 2002; Page & Meyer 1996; 1999). Many towns and cities also function as meeting places, major tourist gateways, accommodation and transportation hubs, and central places to service the needs of visitors. Many tourist trips will contain some experience of an urban area. For example, when an urban dweller departs from a major gateway in a city and arrives at a gateway in another city or region they will stay in accommodation within an urban area. As tourists experience urban tourism in some form during their holiday, to visit friends and relatives, a business trip or a visit for other reasons, it is evident that the welfare of such visitors is a key element in the wider satisfaction with urban areas as tourist destinations (also see chapter 14). Recent research has only belatedly (e.g. Bentley & Page 2001) addressed the concept of tourism as a critical component of visitor satisfaction, of which tourist safety and risk is an integral component.

This chapter examines some of the issues of tourist safety within urban environments and considers the role of special events in potentially influencing the level of safety and risk for visitors to urban areas. The America's Cup hosted in Auckland, New Zealand in 1999/2000 provides an example of how an event may impact upon tourists' perceptions and experiences of tourist safety in an urban environment.

Tourist Safety, Urban Areas and Special Events: The Literature Reviewed

The potential for safety issues to adversely affect travel behaviour and visitor experiences during their holiday or leisure time has resulted in a steady growth in

Managing Tourist Health and Safety in the New Millennium
Copyright © 2003 by Elsevier Science Ltd.
All rights of reproduction in any form reserved.
ISBN: 0-08-044000-2

literature on tourist safety from both academics and public bodies (e.g. World Tourism Organization 1998). Urban areas have the potential to absorb large numbers of visitors in the built environment and the staging of a special event can easily swell these numbers. The hosting of such events may establish an environment that contributes to the concentration of tourism and tourist opportunities for crime. One of the most serious impacts of hosting events is the potential threat to tourist safety (Standeven & De Knop 1999) and this is evidenced by reports of increased crime as one of the negative impacts of special events (Hall *et al.* 1995).

Various research (Fujii & Mak 1980; McPeters & Stronge 1974; Prideaux 1994; Rothman 1978; Walmsley *et al.* 1983) has argued that tourism and crime are related to the extent that patterns of tourism activity coincide with changes in the level of crime. The study of human ecology provides a sociological perspective on crime that has become known among tourism crime researchers as Routine Activities Theory and Hot Spot Theory.

Routine Activities Theory is based on the assumption that predatory crimes feed off the routine activities of others. Cohen & Felson (1979) argue that most criminal acts require the convergence in time and space of a suitable target or victim, a motivated offender and the absence of a guardian capable of preventing the interaction between offender and victim. The suitability of the target or victim is influenced by their value (monetary or symbolic), visibility (e.g. ethnicity, dress and behaviour), access (the offender's ability to access and escape the scene) and inertia (the ability to acquire and dispose of the item of value, or resistance of victim). In the context of tourism, these factors are often pronounced such that tourists may possess a more favourable ratio of risk to reward than locals.

Hot Spot Theory argues that predatory crime is associated with certain types of geographical areas, such that relatively few locations or hotspots are associated with a high percentage of crimes (Schiebler *et al.* 1996). Many of these hotspots exist in urban areas. Crimes against tourists are likely to cluster in these areas involving the concentration of tourism amenities and attractions, and therefore by implication, are likely to be higher in areas hosting special events.

It is not just the location of visitors within cities that predetermines the spatial characteristics of tourist crime. Tourists and tourist areas possess a range of characteristics that make them vulnerable to crime (Chesney-Lind & Lind 1986) and tourists may have a statistically higher chance of being victimised than residents (de Albuquerque & McElroy 1999; Fujii & Mak 1980). A transient population comprising local, domestic and international visitors increases the potential targets for crime and the individual anonymity for offenders. This is combined with the tendency for some tourists to lower their security consciousness on holiday, indulge in risk taking behaviour and enter unfamiliar environments which increases their exposure to criminal activity.

The study of crime in the tourism literature has focused on tourists as victims of crime (Allen 1999; Barker 2000; Chesney-Lind & Lind 1986; Kelly 1993; Schiebler *et al.* 1996; Walmsley *et al.* 1983). A common shortcoming of these studies, with the exception of Allen (1999) and Barker (2000), has been the inability to identify the characteristics of the tourist as a victim, or to differentiate between tourists and residents. Chesney-Lind & Lind (1986) and de Albuquerque & McElroy (1999) differentiate tourists from residents, but were unable to identify any specific characteristics of victims. So how do researchers collate evidence on tourist crime?

The measurement of crime as an impact is based on criminal records of arrest, offence and occasionally victim data, although such statistics are often fraught with measurement problems as not all crimes are reported or recorded. The empirical study of tourist victimization is significantly underrepresented in the tourism literature, as it remains a new and sensitive area of research in many destinations. Tourism authorities are reluctant to measure and potentially risk disclosing the level of crime because of the threat it poses to future visitor numbers (Schiebler *et al.* 1996).

Related research (Burns & Mules 1989; Hall *et al.* 1995; Kelly 1993) has argued that increases in criminal activity accompany the hosting of special events. One of the most widely researched events related to crime was the 1987 America's Cup in Fremantle, Australia. Hall *et al.* (1995: 37) found that the "evidence of the impacts of hosting the America's Cup on criminal and illegal activity in the Fremantle area is substantial". A correlation between the hosting of the event and an increase in criminal activity, particularly for alcohol-related offences, was reported (Hall *et al.* 1995; Selwood & Hall 1988). Selwood & Hall (1988) also found an explicit relationship between the hosting of the 1987 America's Cup and an increase in petty crime. Increases in major personal crimes of sexual and common assault and robbery occurred together with significant increases in minor offences (e.g. traffic infringements, drunkenness and disorderly behaviour) (Hall *et al.* 1995). But how do these findings on the 1987 America's Cup relate to more recent experiences of hosting the America's Cup?

The 1999/2000 America's Cup in Auckland: Developing Data Sources on Tourist Crime

New Zealand first challenged for the America's Cup in Fremantle in 1987, but it was not until 1995 that Team New Zealand successfully challenged for the Cup and won the rights to host its thirtieth defence in Auckland. The Auckland region has a population of just over one million residents and the city receives around 73% of the 1.6 million international visitors to New Zealand annually. Hosting special events is a major focus for the city's tourism organization, Tourism Auckland.

Primary data for the study of crime and the America's Cup was obtained from a random survey of domestic and international visitors to Auckland. Surveys were conducted over an 11 week period between December 1999 and the conclusion of the America's Cup Regatta on 2 March 2000. Personal interviews of 1,003 non-resident visitors were conducted in downtown Auckland and the Viaduct Basin where the America's Cup syndicates were based and a 'Cup Village' was constructed for the event (see Plate 13.1). The survey aimed to identify visitors' perceptions, concerns and experiences of crime during the America's Cup event. The resulting data enabled the victimization probabilities of different tourists to be calculated, while regression analysis was used to predict differences between domestic and international tourists based on details of the offence.

Plate 13.1.: The America's Cup Village, Auckland is an enclosed environment
with a limited number of entry and exit points to reduce opportunistic crime
(Copyright S. J. Page).

A second method of data collection involved the use of tourist victim information reports (TVIR's), a technique previously applied by Barker (2000). Similar to police offence reports (OR's) used to record crime in the general population, the victim reports were designed specifically to record offences committed against domestic and international tourists. The recording of victim data was undertaken by the New Zealand Police after permission and consultation to derive crime data, and the survey was conducted during December to March. However, the tourist-victim data applies only to offences reported to three police stations in central Auckland (i.e. Auckland Central, Downtown and Wharf stations — the latter being inclusive of the America's Cup Village). Tourist reporting of crime at other Auckland stations would have been insufficient to justify the additional resources necessary to record crimes during the America's Cup based on senior police experience of tourism and crime in the city during events. One of the recurring problems with studying tourist crime to date has been the lack of data on the characteristics of tourists as victims of crime (de Albuquerque & McElroy 1999; Prideaux 1994). A major objective of the victim reports was to record details of both the offence and the victim. The ability to develop a detailed profile of tourist crime enabled the study to identify the nature of tourist crime and the characteristics of tourists as victims. The victim reports used in this study represent a mutual and positive relationship with police as part of a proactive approach towards understanding and monitoring future developments in tourism-related crime.

Auckland Police provided secondary data for the study in the form of official statistics extracted from their database monitoring crime in the city. This data included area crime statistics for the Auckland district and the total arrests made by the Operation Marlin unit during the policing period of the America's Cup. Operation Marlin was a once only policing initiative developed for the America's Cup. The operation involved an additional 182 police and was based on similar operations at previous special events such as Asia Pacific Economic Community in Auckland and the America's Cup in Fremantle. Due to the differences in how offences come to the attention of police and are subsequently represented in the data, arrest offences, by nature, predominantly include crimes by offenders that are detected by police (e.g. disorderly behaviour), whereas offence reports reflect crimes reported to police by victims (e.g. theft).

Baseline Data: Auckland Police Area Crime Statistics

Auckland city crime statistics were obtained over a four-year period for the months October to February, which coincided with the period of the America's Cup. Table 13.1

Table 13.1: Auckland City crime statistics October 1999–February 2000.

Offence (Oct–Feb)	1996/ 1997	% change	1997/ 1998	% change	1998/ 1999	% change	1999/ 2000
Assault	331	1.5	336	−6.5	314	16.2	365
Sexual	54	−38.9	33	−9.1	30	16.7	35
Drugs (Cannabis)	337	−3.3	326	12.9	368	6.3	391
Disorder	532	−0.6	529	27.2	673	11.9	753
Burglary	607	−8.1	558	−23.3	428	26.4	541
Theft	2373	−26.6	1741	19.0	2072	0.5	2082
Theft ex car	1409	−37.6	879	30.9	1151	−15.2	976
Total*	27226	−10.5	24354	−5.2	23081	3.3	23837

* Includes other crimes not specified within this table
(New Zealand Police).

indicates that for the period from October 1999 to February 2000, all listed offences experienced an increase compared with the previous year (particularly burglary and assaults) with the exception of thefts from vehicles, which declined. Drug and disorderly offences during the event increased by a lower margin than in the previous year, although it is likely that the recording of these crimes were substantially underestimated based on police tolerance towards these offences. While total crime levels in Auckland city increased by 3.3% during the America's Cup, Auckland's population experienced a 9.5% rise in international visitors to New Zealand during this period (Statistics New Zealand 1999; 2000). Corresponding visitor statistics to central Auckland were unavailable. Crime statistics for the period 2–5 March 2000 magnify the hedonistic impacts of a special event on crime over a concentrated period of time, where this period represented the end of the America's Cup and the celebrations that endured into the weekend. The concentrated impact of the America's Cup was reflected by a high incidence of behaviour, assault and willful damage offences, as might be expected given the combination of day and night celebrations, large crowds and alcohol consumption. Of further note were the 50 recorded cases of theft from vehicles, which again reflected the large crowds and increased opportunities for crime.

To consider the temporal variation in the pattern of these crime statistics a correspondence analysis (Greenacre 1984) was completed. Correspondence analysis allows the associations between the rows and columns of a frequency table to be illustrated in a plot that suggests the proximity of the row and column categories. Such plots are particularly useful when the large number of categories makes a cross-tabulation difficult to interpret, as in this case. Some caution is advisable when interpreting correspondence analysis plots, especially when the association between the rows and columns is insignificant. Fortunately, this is not of concern in this example because the sample size is large ($N = 20,224$, $\chi^2 = 195$, $DF = 18$, $p = 0.000$). However, in any correspondence analysis plot, the plot for the rows is never strictly comparable with the column plot, so the proximity of row and column points can only be approximately measured by creating axes from crucial points to the origin. For example, the axis created for sexual crimes in Figure 13.1 suggests that the proportion of sexual crimes has fallen steadily from 1996/1997 to 1999/2000. Additionally, the axis for disorder suggests that the proportion of disorder offences increased steadily from 1996/1997 to 1999/2000. In this respect it appears that the America's Cup has not changed existing trends.

The plot explains a very healthy 98.6% of the association between year and type of crime, with the horizontal axis explaining a dominant 68.4% of the association. This horizontal axis positions 1996/1997 on the far right and 1999/2000 on the far left so it has been labelled 'Time: Present to Past'. The vertical axis puts violent crimes (e.g. burglary, sexual and assault) at the bottom and less violent crimes (e.g. disorder and theft ex car) at the top so it has been labelled 'Level of Crime: Violence to Disorder'. The position of 1996/1997 and 1999/2000 towards the middle of this vertical axis suggests that these years were similar in terms of the level of violent crime, while 1997/1998 saw relatively more violence and 1998/1999 saw relatively less violence.

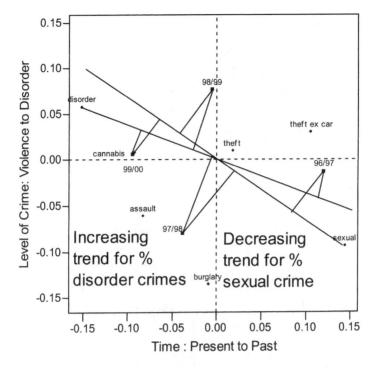

Figure 13.1: Trends in the Auckland City crime statistics.

Auckland Police Arrest Statistics (Operation Marlin)

Policing of the America's Cup by Operation Marlin involved a geographical area extending from the Auckland Harbour Bridge to Quay Street Reef including the Viaduct Basin (see Plate 13.2) and America's Cup Village, and covered what might be described as within the city's main tourist district which extends to the waterfront area following the construction of the America's Cup Village. Between October 1999 and early March 2000, Operation Marlin police arrested 511 persons for a total of 745 offences (Figure 13.2). Despite more than four million visits to the Cup Village in this period, only 33.1% of these arrests took place within the Operation Marlin policing.

Drug and anti-social offences collectively accounted for 45.1% of all offences and the highest proportion of arrests. Of these, 23.1% of offences comprised drunk and disorderly behaviour, while drug offences accounted for 14.5% of crime. The high incidence of disorderly behaviour was in part, a reflection of the increased detection of these crimes by police, the easy availability of alcohol, and the tolerance by licensed premises and police towards alcohol consumption, especially outdoors and in public places at night.

Crimes of violence comprised 20.7% of all detected offences, of which 6.7% were for assaults (excluding sexual assault) against members of the public. An assumed correlation between crimes of violence and night-time activities, including the

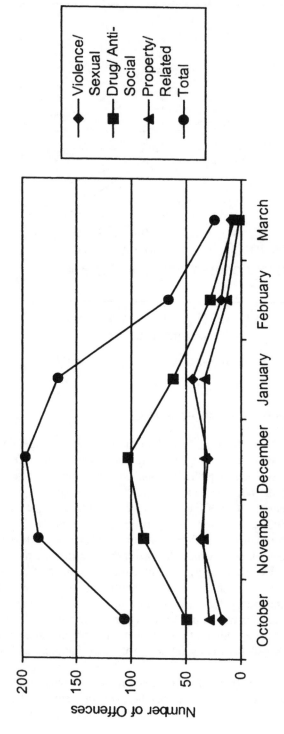

Figure 13.2: Operation Marlin arrest offences October 1999–March 2000.

Source: New Zealand Police.
March figures represent an incomplete month.

consumption of alcohol, is presumed to account for a significant proportion of these crimes. Property and related offences accounted for 19.3% of offences and the remaining 14.9% comprised traffic offences. The decline in the number of arrests from 197 in December to 66 in February is explained by the end of the Christmas and Millennium celebrations, the return to work by many in the domestic population, and a resulting reduction in the number of Operation Marlin police at the end of the Louis Vuitton Cup (challenger series) in the first week of February. By comparison, the TVIR data shows that reported offences by tourists were more consistent over this period, with 50 offences reported in December, 80 in January and 68 in February.

Tourist Victim Information Reports

A lack of data on the victim characteristics of tourists became a reason for using the TVIR's, which sought to differentiate tourists from the aggregated official crime data. Victim reports only provide a measure of crimes against tourists that were reported to police and cannot be assumed to represent the characteristics of unreported crime. Between December 1999 and March 2000, visitors to Auckland reported 202 criminal offences to police, of which 43 (21.6%) were reported by domestic tourists and 156

Plate 13.2.: The America's Cup Village, Auckland illustrating the creation of a mixed residential and cafe/hospitality development planned and enclosed to create a living and working tourism district that is ambient and a visitor destination within the city (Copyright S. J. Page).

(78.4%) by visitors from overseas. European tourists accounted for the highest incidence of victimization (26.3%), while Japanese and other Asian tourists comprised 7.8% of victims collectively, reflecting the visitation rates from these regions. Over half (71.4%) of all victims were 20–29 years of age and 67.8% were male, which supports the traditionally higher risk and risk-taking behaviour of young males noted in the criminological literature (Ongley 1996). A predominance of property crimes (98.7%) over violent crimes (1.7%) was found, whereby theft from vehicles accounted for 55.4% of reported offences and other theft comprised 39.1%.

One of the major differences in tourist victimization rates is based on the domestic or international origin of the tourist and their respective exposure to risk. Overseas tourists were more likely to experience theft from accommodation or person than domestic tourists, reflecting their greater tendency to use commercial accommodation and campervans. Domestic tourists were more likely to experience crime in public places, particularly theft from vehicles, which reflects the greater self-drive travel among this group. In terms of location, public places (55%), accommodation (15.8%) and campervans (10.4%) accounted for the highest share of all crimes respectively. Overseas visitors also incurred a far greater monetary loss for their possessions than domestic tourists since the losses from overseas tourists' accommodation and campervans involved greater quantities of items than domestic tourists' losses from theft from vehicles.

Auckland Visitor Survey of Tourist-Related Crime

The sample of 1,003 visitors to Auckland for the visitor survey comprised 29.3% of domestic tourists and 70.7% of tourists from overseas. The nationalities of international visitors in the sample encompassed 45 different countries including a significant proportion of European and North American visitors, while less than 10% of visitors were of Japanese or other Asian decent. A low proportion of Asian visitors in the vicinity of the Viaduct Basin was also noted during the Whitbread stopover in 1993/1994 (see Burgess & Molloy 1994). An interest in yachting and a high proportion of yachts competing in the America's Cup with the European and American syndicates were likely reasons for the sample representation of visitors. Although the existing composition of visitors to Auckland was likely to change during the America's Cup, no comparable data that recorded the origins of visitors to the city during that time was available. Using an English language survey and a lower willingness to participate in the survey among Asian and Japanese visitors also affected the representation for groups in the sample.

The survey asked visitors whether they believed that they had been criminally victimised in one or more of the five crime experiences. Some 30 respondents, or 3% of the sample, reported a total of 34 offences with multiple responses allowing for the fact that any one person can be victimised on more than one occasion. Of the total offences reported in the survey, 50% involved the theft or burglary of property from tourists' accommodation and 29.4% involved theft from vehicles. Only two incidents of violence against tourists were reported, which accounted for 5% of the total offences.

Victim Characteristics of Tourists

Perhaps the primary benefit of recording the characteristics of tourist victims of crime is the ability to identify differences in the risks among different tourists and travel options. Overseas tourists for instance, represented 70.7% of the sample, yet they accounted for 80% of victims ($\pm 15\%$ with a 95% confidence). This is an interesting but not a statistically significant difference. The highest levels of reported victimization against visitors were from the United Kingdom (20.0%), Australia and Europe (16.7%). Non-New Zealanders were disproportionately more likely to be the victims of crime, whereby 3.8% of these tourists were victimized compared with 1.3% of tourists citing their nationality as a New Zealander. In terms of vulnerability, overseas tourists were 1.6 times more at risk of crime than domestic tourists.

The choice of visitor accommodation was a considerable factor in explaining crime, whereby half of all reported crimes occurred at the place of accommodation. Domestic tourists were twice as likely to stay with friends and relatives, while international tourists were more inclined to stay in backpacker hostels, and these differences in exposure to risk influenced the rates of victimisation between these groups. Tourists staying in backpacker hostels experienced the highest level of crime (39.3%), followed by the accommodation choices of friends and relatives (32.1%) and camping or campervans (17.9%). The offences did not always occur at these locations. Thus, in terms of the highest rates of victimisation, 8.1% of tourists who camped or stayed in campervans and 7.2% of those who stayed in backpacker hotels were the victims of crime. Furthermore, of those crimes that occurred in places of accommodation, 58.8% occurred in backpacker hostels, reflecting the lower security offered at these premises located in downtown Auckland.

Despite comprising less than half (47.2%) of all respondents, 73.4% of crime victims were aged between 20 and 39 years. Visitors aged less than 40 years were over-represented as victims of crime while those aged 40 years and above were under-represented. A decreasing rate of victimisation with increased age is consistent with both the victim reports and the Bureau of Crime and Statistical Research study findings (Allen 1999). The size of the travel group was also important in explaining victimisation rates among tourists whereby 55.2% of crime victims indicated that they were travelling alone.

Evaluating the Tourism-Crime Nexus and the America's Cup

Relatively low rates of victimization were recorded at the America's Cup, reflecting the comparatively safe image of New Zealand for visitor activity. This was confirmed by detailed discussions with senior police. Crime increased during the America's Cup as expected with an increase in population, however at a level that was less than the proportional change in population. The proportion of property to personal crime supports the predominance of property crimes experienced by tourists, and the value and access associated with tourists' possessions. Arrests for violent offences accounted for one-fifth of offences, although the low rate of reported crimes against the person reflected the differences in reported and detected crime, and low reporting rates for personal crimes. The increase in crime during the Cup was undoubtedly an underestimate of the true extent of crime due to a police tolerance towards minor offences and under-reporting of crime, particularly among tourists who typically have lower reporting rates than residents. The nature of reported crime against tourists cannot, therefore, be assumed as a reliable basis for predicting the level of unreported crime.

One of the major difficulties with understanding event related crime is in ascertaining the proportion of crime attributable to the increase in population, the increase in tourism activity and that associated with hosting the event. Table 13.2 depicts the major event and destination-based variables associated with an increased visitor population that were considered fundamental in determining the low incidence of crime during the America's Cup in Auckland, and whereby the rating of each variable along a continuum corresponds to an associated level of crime.

The nature of special events is related to the impact of crime based on the scale of the event, the impact on the host community and the marketing focus. Small scale events for instance, are less likely to be associated with negative problems like crime compared to large-scale or mega events (Hall 1992). Crime is influenced by the levels of social and hedonistic activity at events, particularly at night, and can lead to increases in alcohol and drug-related offences, while large crowds can be conducive to pickpockets and snatch thefts. Prideaux (1996) argued that the promotion of a tourism locality as a hedonistic destination attracts a type of visitor and associated activity where rates of crime may increase.

The presence and visibility of police or 'guardians' has been reported for their effect as a deterrent to crime in tourist destinations by increasing offenders' risk of apprehension (Cohen & Felson 1979; Jud 1975; Kelly 1993; McPeters & Stronge 1974; Pizam *et al.* 1997; Prideaux 1994; Rothman 1978).

Urban tourism environments can create hotspots of criminal opportunities due to the influx of tourists and workers, increased crowding and anonymity, and enclaves of

Table 13.2: Variables related to crime at special events.

	Low Crime		High Crime
Event type	Family	- - - - - - - - -	Hedonistic
Destination type	Leisure-oriented	- - - - - - - - -	Tourism-oriented
Visitor type	Family	- - - - - - - - -	Hedonistic
Community support	Majority	- - - - - - - - -	Minority
Overall event impacts	Positive	- - - - - - - - -	Negative
Event duration	Long	- - - - - - - - -	Short
Police presence	High	- - - - - - - - -	Low
Location/spatiality of event	Concentrated	- - - - - - - - -	Spread
Existing crime levels	Low	- - - - - - - - -	High
History of crime at event	Low	- - - - - - - - -	High
Media profile	Low	- - - - - - - - -	High

accommodation, attractions and entertainment (Fujii & Mak 1979; Kelly 1993; Prideaux 1994; Richter & Waugh 1986; Ryan 1993; Schiebler *et al.* 1996). Such locations can also create a safe enclave for tourists if there is sufficient police and security presence to deter crime, as demonstrated by Operation Marlin. The environmental design of the America's Cup Village limited the number of access points to seven and particular attention was paid to lighting and a visible, yet unobtrusive, presence of police and security. In this way, the nature of the environment, visitors and the event substantially reduced the capacity for offenders to successfully engage in criminal activities.

An enclave area like the America's Cup Village can lead to issues of displacement of crime such that the greatest risk of criminal victimization was not in the immediate areas confined by America's Cup activities, but as visitors returned to vehicles, accommodation or visited other attractions of the city where security was less concentrated. Available police data confirmed that only 169 arrests took place in the America's Cup Village in almost five months. The relocation of additional police resources in the America's Cup area may provide opportunities for crime at these source areas, where the opportunity and risk were more favourable. Auckland Police argue that this form of crime displacement was minimised because the personnel assigned to police the America's Cup were in excess of existing district quotas.

The impact of crime during special events is also dependent on the level of existing crime occurring in the wider host destination and encompasses significant commercial and political risks for special events embroiled in negative publicity. Schiebler *et al.* (1996) argue that tourist crime is more likely to occur in destinations already experiencing high rates of crime, and introducing tourism to a destination with low crime will not invoke an increase in crimes against tourists in the same way as a high crime destination. In this way, crime was more of a potential concern than an existing concern in Auckland, which had experienced an 11.6% decline in recorded crime in the years to December between 1997 and 1999 (Auckland City Police and Auckland City District Crime Statistics 2000).

Conclusion

It is evident that urban destinations are likely to have hotspots of criminal activity which tourists are advised to avoid. When the impact of a special event is considered, it is likely that some degree of additional crime will be generated by the event. However, what the Auckland America's Cup suggests is that there is no consistent model or rate of crime, which can be attributed to such an event. The increase will partly depend upon the nature of crime already endemic in the destination. The characteristics of the visitors, their activity patterns, behaviours and willingness to be vigilant in an unknown destination will also be important factors to consider. What is problematic is gauging the full scale, extent and nature of tourism related crime. It inevitably requires a triangulation of numerous data sources and cooperation with policing bodies to gather relevant information.

Crime at special events is also dependent on a range of interrelated variables associated with the event and the host destination, as well as the status of the population and seasonal increases in tourism and crime. Yet, in the case of the America's Cup in Auckland, these variables were more favourable in deterring crime than increasing crime at a rate greater than the proportional increase in population. Thus, the increase in crime attributable to the America's Cup was low and less than the change in visitor population at risk. Examination of the research data reveals that differences in ethnicity, accommodation choice and to a lesser extent the age of visitors to an event destination affect their risk of criminal victimisation. This is reflected in the victimisation rates between international tourists and domestic tourists, and the resulting criminal victimization is based on the exposure to risk that these factors entail. Overseas tourists were more likely to be the victims of thefts from accommodation, while thefts from vehicles were higher among domestic tourists. Although there were notable differences in the nature of crimes against domestic and overseas tourists, differences in the actual victimisation rates were not significant.

There is also intrinsic value in collecting statistics on tourism and crime from a practical perspective and the value of close collaboration with the police to identify changes in the nature and level of crime associated with a special event. It also indicates that irrespective of the level of crime, tourism and security officials can utilise this data to plan for events and enable effective use of the limited resources available to police. The co-ordination of the timing and planning for events between the tourism sector and police is strongly advocated by the World Tourism Organisation (1998). Yet, as with all forms of tourism data, longitudinal studies are required to develop more robust models of the relationship between tourism and crime.

References

Allen, J. (1999). *Crime against international tourists*. NSW Bureau of Crime Statistics and Research. Available at www.lawlink.nsw.gov.au/bocsar/

Auckland City Police and Auckland City District Crime Statistics (2000). Available at http://www.auckland_citypolice.govt.nz/Facts_and_Figures/crime_statistics.htm

Barker, M. (2000). *An empirical investigation of tourist crime in New Zealand: Perceptions, victimisation and future implications*, Unpublished doctoral thesis, Centre for Tourism, University of Otago, Dunedin, New Zealand.

Bentley, T. A., & Page, S. J. (2001). Scoping the extent of tourist accidents in New Zealand. *Annals of Tourism Research, 28* (3), 705–726.

Burgess, C., & Molloy, T. (1994). *1993–1994 Auckland Whitbread stopover: Economic impact report*. Auckland: Price Waterhouse and Massey University.

Burns, J. P. A., & Mules, T. J. (1989). An economic evaluation of the Adelaide Grand Prix, In: G. Syme, B. Shaw, D. Fenton, & W. Mueller (Eds), *The planning and evaluation of hallmark events* (pp. 172–185). Aldershot, Gower Publishing Company.

Chesney-Lind, M., & Lind, I. Y. (1986). Visitors as victims: Crimes against tourists in Hawaii. *Annals of Tourism Research, 13*, 167–191.

Cohen, L. E., & Felson, M. (1979). Social change and crime rate trends: A routine activity approach. *American Sociological Review, 44*, 588–608.

de Albuquerque, K., & McElroy, J. (1999). Tourism and crime in the Caribbean. *Annals of Tourism Research, 26* (4), 968–984.

Fujii, E., & Mak, J. (1980). Tourism and crime: Implications for regional development policy. *Regional Studies, 14*, 27–36.

Greenacre, M. (1984). *Theory and application of correspondence analysis.* London: Academic Press.

Hall, C. M. (1992). *Hallmark tourist events: Impacts, management and planning,* London: Belhaven Press.

Hall, C. M., Selwood, J., & McKewon, E. (1995). Hedonists, ladies and larrikins: Crime, prostitution and the 1987 America's Cup. *Visions in Leisure and Business, 14* (3), 28–51.

Jud, G. D. (1975). Tourism and crime in Mexico. *Social Science Quarterly, 56*, 324–330.

Kelly, I. (1993). Tourist destination crime rates: An examination of Cairns and the Gold Coast, Australia. *Journal of Tourism Studies, 4* (2), 2–11.

McPeters, L. R., & Stronge, W. B. (1974). Crime as an environmental externality of Tourism: Miami, Florida. *Land Economics*, 288–292.

Ongley, P. J. (1996). *Crime.* Wellington, New Zealand: Statistics New Zealand.

Page, C., & Meyer, D. (1999). *Applied research design for business and management.* Sydney: McGraw-Hill.

Page, S. J. (1995). *Urban tourism.* London: Routledge.

Page, S. J., & Hall, C. M. (2002). *Managing urban tourism.* Harlow: Pearson Education.

Page, S. J., & Meyer, D. (1996). Tourist accidents: An exploratory analysis. *Annals of Tourism Research, 23* (3), 666–690.

Pizam, A., Tarlow, P. E., & Bloom, J. (1997). Making tourists feel safe: Whose responsibility is it? *Journal of Travel Research, 36* (1), 23–28.

Prideaux, B. (1994). Mass tourism and crime: Is there a connection? A study of crime in major Queensland tourism destinations. In: *Tourism research and education conference* (pp. 251–260). Brisbane: Bureau of Tourism Research.

Prideaux, B. (1996). The tourism crime cycle: A beach destination case study. In: A. Pizam, & Y. Mansfeld (Eds), *Tourism, crime and international security issues* (pp. 59–76). Chichester: John Wiley & Sons.

Richter, L. K., & Waugh, W. L. (1986). Terrorism and tourism as logical companions. *Tourism Management, 3*, 230–238.

Rothman, R. A. (1978). Residents and transients: Community reaction to seasonal visitors. *Journal of Travel Research, 16*, 8–13.

Ryan, C. (1993). Crime, violence, terrorism and tourism: An accidental or intrinsic relationship? *Tourism Management, 14*, 173–183.

Schiebler, S. A., Crotts, J. C., & Hollinger, R. C. (1996). Florida tourists' vulnerability to crime. In: A. Pizam, & Y. Mansfeld (Eds), *Tourism, crime and international security issues* (pp. 37–50). Chichester: John Wiley & Sons.

Selwood, H. J., & Hall, C. M. (1988). *The hidden underbelly: Some observations on the unpublicised impacts of the America's Cup.* Canadian Association of Geographer's Annual Conference, St. Mary's University, Halifax.

Standeven, J., & De Knop, P. (1999). *Sport tourism.* Champaign, Illinois: Human Kinetics.

Statistics New Zealand (1999). *International visitor arrivals to New Zealand.* Wellington: Statistics New Zealand.

Statistics New Zealand (2000). *International visitor arrivals to New Zealand.* Wellington: Statistics New Zealand.

Walmsley, D. J., Boskovic, R. M., & Pigram, J. J. (1983). Tourism and crime: An Australian perspective. *Journal of Leisure Research, 15* (2), 136–155.

World Tourism Organization (1998). *Tourism: 2020 vision – executive summary.* Madrid: World Tourism Organization.

Conclusions

Conclusion

Chapter 14

Risks, Rights and Responsibilities in Tourist Well-Being: Who Should Manage Visitor Well-being at the Destination?

Linda Walker and Stephen J. Page

Introduction

In many developed countries, governments have shown a growing interest in improving the well-being of the community (Grant 2002). Policy documents and initiatives have been developed such as 'community safety' with the aim of enhancing the well-being of the community. Tourism can be seen as a tool to further the aims of community well-being through benefits such as providing employment opportunities, acting as a catalyst to protecting the environment or increasing the viability of local facilities. The arguments regarding the benefits of tourism to the local community are well documented (Page & Hall 2002). However, it should be noted that despite some dissident voices in the academic community on the true value of tourism to a locale, there are few countries, regions or localities that do not actively encourage tourism development. Most governments, including the U.K. (and the Scottish Parliament) believe tourism to be beneficial to local communities. However, there has been little empirical research into the well-being of the tourist while visiting the communities that act as destinations during the tourist visit; this despite a growing interest in the way the visitor is handled, accommodated and managed in the destination in terms of their well-being. This chapter examines the differences in terms of risk between residents and visitors before examining the debates associated with the reasons for taking responsibility for visitor safety and security. It concludes with a discussion on the allocation of responsibilities within the destination which will have policy and implementation issues for the tourism and non-tourism industry sectors and agencies.

Prior to discussing these issues, it is pertinent to review the concept of tourist well-being, based on the studies published by Page & Meyer (1996), Page (1997) and Bentley & Page (2001).

Tourist Well-Being: A New Concept for the Tourism Industry

As discussed in Chapter 1, the term 'tourist safety' equates to concerns for the well-being, welfare and wider safety of the visitor not only while travelling from the origin area to the destination area, but particularly the way in which their personal safety is affected by their activity patterns in their own action space. Many new tourist attractions are developing environments in which visitor safety is paramount, often in consultation with the local police forces. For example, Plate 14.1 from Lomond Shores, a new mixed development of retailing and a large visitor attraction/visitor centre on the banks of Loch Lomond in Scotland illustrates many of these principles at work. Plate 14.2 shows that the environment is a self-contained development with a limited number of entry and exit points which are under surveillance from CCTV (Plate 14.3) and clearly visible areas with unobscured views and well-lit car parking areas (Plate 14.4) have been provided. This is designed to deter crime (e.g. break-ins to cars) as well as vandalism as well as reassuring visitors of the safety and security measures built-in to the retailing environment. This raises notions of well-being and satisfaction which few studies seem willing to acknowledge because it requires a holistic assessment of the tourist event. As chapters throughout this book have emphasised, one of the problems of developing a wider conceptualisation of tourist safety means we need to move towards a more holistic

Plate 14.1: Lomond Shores is a £60 million urban regeneration project designed to create a gateway to the new Loch Lomond and Trossachs National Park in Scotland, based on a mixed development of a visitor attraction, upmarket retailing and a gateway visitor centre (Copyright S. J. Page).

assessment where a new research paradigm, based on well-being, is necessary. This places the tourist as the focal point of the research and requires a wide range of multidisciplinary skills to assess the wider context of safety issues and how they impact on tourists' well-being. It is no longer acceptable for tourism managers and the wider tourism industry to view the tourist as a passive consumer who is unaffected by the use and interface with tourist health and safety-related services. Just as the tourism sector now actively uses marketing to position destinations and products based on unique selling propositions (USPs), then there is a growing role for building in these elements of well-being, particularly health and safety in the imaging of destinations and specific tourism products. At a purely commercial level, this gives destinations and businesses a competitive edge over those destinations which do not attempt to build in notions of caring for a visitor's well-being, given the world-wide concerns for tourist safety post September 11th and the Bali bombing. Indeed, the tourist is now seen as a legitimate target for attack and assuring visitors that their well-being is being actively safeguarded, looked after and built into the management of a destination will certainly pay dividends for those destinations seeking to position themselves as desirable. The effects of terrorism and indiscriminate attacks on the tourism sector are very evident in destinations within Israel that has seen its industry plunged into crisis during the last two years. Images, visitor perception of their well-being and particularly safety have become paramount in the new millennium, beyond anything that analysts could have forecast.

Plate 14.2: The design of the spatial layout of the development as shown on this visitor information board highlights the limited number of entry and exit points to reduce opportunist crime and to enhance visitor safety (Copyright S. J. Page).

Plate 14.3: CCTV is a visible element of the crime-deterrents in place together
with anti-ram raiding elements to reduce planned retail-related crime
(Copyright S. J. Page).

Many of the chapters in this book have shown that a disparate range of professions
interact with tourists over issues such as health and safety and this may be infrequently,
in the case of litigation by tourists, through to health professionals on a more regular
basis. It is also important to recognise the macro issues affecting destinations such as
policy making which can have an enormous bearing on the image of how well-being is
accommodated in tourism. For example, in destinations the policy decisions on whether
the tourism industry is highly regulated, is self-regulating through to unregulated will
have a significant bearing on tourist well-being, especially if things go wrong or service
interruptions occur due to poor co-ordination and planning by the public sector. As

emphasised in Chapter 1, in a tourism context, this perspective can be applied to understand how their well-being is conditioned and affected by what they do, where they stay and the risks they are exposed to. In other words, one needs to understand the interaction of the tourist and tourism industry in a particular locale and how the safety issues impact upon their well-being. Yet the conceptualisation, analysis and discussion of the factors that may negatively or positively impact upon the tourist well-being are diverse. These factors need to be viewed against their dynamic role as transient elements in destination areas that arrive and depart and do not have the same stakeholder role in the area as residents and businesses.

The Tourist as a Transient Visitor in Destinations: The Scope of the Impacts and Issues for Destinations

Tourist normally utilise the services and facilities of the community that they are visiting. However in terms of non-commercial services and facilities such as policing and emergency services, the incoming tourist may be utilising a service developed for the local community but it may not be developed with that group in mind. While this may not prove problematic in the early stages of tourism development, it becomes

Plate 14.4: Visible and well-lit car parking areas are a hallmark of environments which seek to design crime out and integrate safety in to the spatial layout of visitor activity (Copyright S. J. Page).

inappropriate for larger scale developments. The main problems which tourist growth poses for health and safety includes:

- *Lack of awareness* — visitors may be unaware of the services and facilities available to them;
- *Over-stretched resources* — the volume of visitors, particularly in a very seasonal destination, may strain the services and facilities already in place;
- *Resentment* — the community may feel aggrieved if they feel that visitors are using services and facilities at the expense of the local residents;
- *Differing needs* — the visitor may not be subject to the same type of incidents as the local residents and may require different services and facilities both for reactive measures and in any preventative measures undertaken.

These factors may result in gaps in provision for the visitor with no single body or organisation taking responsibility to ensure that adequate health and safety provision is made for the influx of visitors to the destination.

Acknowledgement, Acceptance and Allocation

In order for resources to be provided that meet the requirements of the visitor there must be an *acknowledgement* that tourists have distinct needs; an *acceptance* that the community/destination should provide for these needs; and an *allocation* mechanism that delivers resources to meet the needs of the visitors (see Figure 14.1). This requires a longer-term strategic view to be taken of the area as a tourist destination that allows decisions to be made at policy level.

The level of services at destination will be dependent on a number of factors, such as the level and adequacy of provision for the local population, how much additional demand is placed on services by the visitor and the priority given to visitors by the

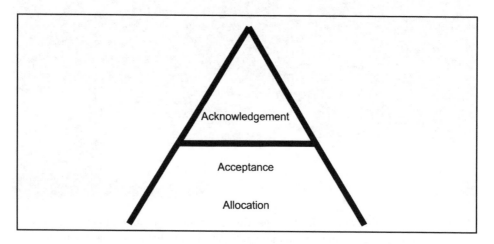

Figure 14.1: The three 'A' model for resource provision.

service providers. Many of the services required by the visitor during their stay are provided for by government at various levels (local, regional, national and supra-national[1]). This means that the well-being of the visitor at destination may, to a great extent, be dependent on government policy on tourism and the value placed on it.

The visitor is the consumer of the tourism product and the destination is where the tourism product is consumed. With several notable exceptions (such as Disney World or Oasis), destinations are not commercially owned and operated entities where there is a clear 'duty of care' with one owner responsible for the safety and security of visitor while at the destination site. The majority of destinations are areas that combine specific tourism and non-tourism businesses within a normal place of residence.

Acknowledgement: Why Does Visitor Well-Being Need to be managed?

Differences in Risk for Visitors and Residents

Destinations can be seen as the focal point of the activities of the tourist/visitor but they are rarely just tourism enclaves. More often the destination is first and foremost a residential area with facilities and services in place primarily for the safety and security of the local population, used by the visitor during their stay. Whether visitors or residents, there is the potential (i.e. the risk of) of becoming involved in an accident or victim of a crime. However, a visitor to an area may be more prone to certain types of incidents occurring and may require different services and facilities to prevent or deal with such an occurrence. For example, many chapters in this book have highlighted the prevalence of certain risk factors and behaviours for visitors and their greater propensity to experience adverse events that can, potentially, damage the tourist experience.

Incidents take many different forms from fairly minor incidents such as trips and falls through to major incidents such as terrorist attacks; they can be seen in the form of a continuum such as that developed by Page (see Figure 14.2); they may reflect particular hazards of the destination such as changeable weather conditions, thieves targeting visitor cars at remote car parks or attacks by terrorist groups. However the visitor, for example due to an existing medical condition, driving without due care or not securing valuables, may also induce incidents. Whatever the cause, the visitor will find themselves subject to the conditions prevalent at the destination in terms of emergency services, health care and other support services. Their experience will be much dependent on the services and facilities available to them as visitors to the destination.

There have been attempts in the field of geography to measure well-being of the individual by using indicators, both specifically measurable such as income levels and other less tangible measures, to make a 'quality of life' measurement (e.g. Pacione & Gordon 1984; Massam 2002). This method of assessing well-being is not directly transferable to the tourist at destination as there are different criteria and implications

---

[1] In Europe, there may also be a provision at supra-national level based on the policy recommendations and implementation by the EU.

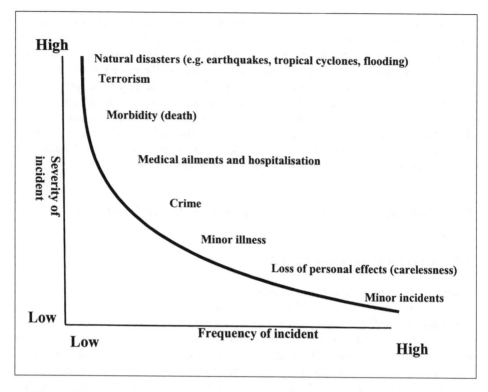

Figure 14.2: The tourist health and safety continuum: severity and frequency of incidents.

for the individual in terms of the both expectation and responsibility. While there are currently no agreed definitions of tourism well-being, it is reasonable to surmise that risks to person in terms of increased susceptibility to crime, accident or illness will affect the well-being of the tourist.

The term destination covers a wide range of typologies; at its simplest level, a destination is the place that tourists (and visitors) visit. However, destinations do vary in type and attributes (see Figure 14.3). In some cases the destination can be totally divorced from the local populace, and normally these 'enclave' destinations are in areas where crime is perceived as a high-risk problem, there may be contributing factors such as a high level of poverty or political reasons for segregating visitor and local. Examples of this 'ghetto' approach to tourism can be seen in countries such as Jamaica, the Dominican Republic and Cuba. Theme parks, traditional holiday camps such as Butlins, or their more recent counterparts, the holiday villages such as Oasis, provide a similar artificial environment where the holidaymakers and visitors are sheltered from the external conditions. This gated community approach to tourism and visitor safety with implementation of surveillance mechanisms, such as CCTV has been discussed by Barker & Page (2002). For 'custom' destinations such as Disney or Oasis, the

Destination Type	Description	Examples	Probable scenarios for health and safety provision			Responsible Agents
			Security/ Policing	Healthcare/ Medical facilities	Environmental risks/ Hazards	
'Enclave' Destination	Segregated tourism facility outwith the normal community	Usually found in areas where crime is perceived a high-risk problem, there is a high level of poverty or there are political reasons for segregating visitor and local. Examples of this 'ghetto' approach to tourism can be seen in countries such as Jamaica, the Dominican Republic and Cuba.	High levels of security provided to 'protect' tourist from external environment	Private health care facilities provided for tourist consumption	Artificial enclosure of site reducing risk from external environmental hazards	Owner of facility Tour operator National government
Custom destination	Purpose-built self-contained destinations with all services and facilities provided within the resort	Commercial destinations such as Center Parcs or Disneyworld	Security mainly controlled by owners/ external policing used where necessary	Some health care provision on site/ external health care agencies used where necessary	Facility owner has control of most aspects of the environment/ must comply with some external regulations	Owner of facility Local service providers
Tailored destination	Built or reinvented with the main purpose of cater for the tourist	Often seaside destinations or former fishing villages such as Blackpool; but can be purpose built or developed around a small original settlement in an area suited to tourism activity such as Aviemore	Policing responsibility mainly lies with local police force but some private security may be employed	Mainly provided by normal health care agencies/ some private health care/ medical centres	Some aspects controlled by site owners/ service providers but mainly controlled and regulated by local authority	Site owners Tour operator Local authority Local service providers
Event destination	A destination only for the duration of the event (may lead to establishment as a destination in its own right)	Large events requiring infrastructure and superstructure to cater specifically for the influx of people for the event such as Greenock Tall Ships event or the Olympic Games	Policing responsibility mainly lies with local police force but private security employed for aspects of security e.g. car parking	Mainly provided by normal health care agencies/ some private health care/ medical centres	Some aspects controlled by event organiser but mainly controlled and regulated by local authority	Event organiser Site owners Local authority Local service providers
Developed destination	A high level of tourism but not designed to cater particularly for their needs	Often towns or cities with particular attraction for tourists and day visitors such as Stirling in Scotland or New York	Policing responsibility lies with local police force but private security employed for aspects of security e.g. car parking	Provided by normal health care agencies	Some aspects controlled by site owners/ service providers but mainly controlled and regulated by local authority	Tour operator Site owners Local authority Local businesses Local service providers
Under developed/ developing destination	A low level of tourism and little provision for specific needs of tourists	Often towns or villages close to a main destination receiving 'spill-over' visitors (mainly day trips) or attempting to develop tourism for example Falkirk in Scotland or Montechoro in Portugal	Policing responsibility lies with local police force	Provided by normal health care agencies	Some aspects controlled by site owners/ service providers but mainly controlled and regulated by local authority	Local authority Site owners Local businesses Local service providers
Latent destination	Very little tourism and little or no attempt to cater for their needs	Outwith the recognised 'tourism areas' but occasionally visited by day trippers or tourists with a particular agenda (VFR/ business)	Policing responsibility lies with local police force	Provided by normal health care agencies	Controlled and regulated by local authority	Local authority Local service providers

Figure 14.3: Probable health & safety scenarios by destination type.

destination services are set up specifically to fulfil visitor needs, being visitor centred. Although there may be a reliance on external agencies to provide some aspects of visitor health and safety, the majority will be dealt with in-house in these environments; the influence of the destination in terms of scale and local income/job provision coupled with control/responsibility by a single organisation, will ensure that the external agencies focus on the needs of the visitor too. There may be an element of this complete focus in other tourist destinations where visitors are recognised as the main income source and vital to the local economy. In this 'tailored' destination there is a higher likelihood that the visitor or their representative (in the form of tourist boards, tour operators or travel agents, for example) will be able to influence the level and type of health and safety provided by the destination. Where this is the case, aspects of the 'custom' destination may occur in the 'developed' and 'tailored' destinations for example in Thailand where they have introduced 'tourist police' and areas such as the Algarve where many health centres are operated specifically for the visitors. Where there are large numbers of package holidaymakers, tour operators often have staff instructed to recommend specific doctors and health centres that will deal with the clients in their own language as individual crises affect the well-being of the holidaymaker. 'Event' destinations are likely to show similar attributes to the 'tailored' destination' with additional health and safety provision put in place for the duration of the event. However, the 'developing' destinations may have less provision for visitors and may rely on the existing local services and facilities. In some cases they may be inadequate even for the needs of the local community and their use by an influx of visitors may cause displacement of services leading to local resentment. Even in a scenario where additional provision is made, this may be at a lower standard than expected by the visitor. While tourism is at a low level or in urban areas where high numbers of visitors can be absorbed without undue strain, the well-being of the visitor is likely to be easily catered for within the existing community structures. However, where there are large influxes of visitors; it is unlikely that the existing structures will have the resources or expertise to cope.

Hazards Into Risks

Both residents and visitors are subject to risk of crime, accidents, illness, injury or even death. However as visitors we may have different levels of awareness of potential hazards and therefore possible risks. Sharp (2001) differentiates hazards from risks as follows:

> "Hazards are defined in absolute terms (e.g. cliff faces, avalanche prone slopes, fast moving water, electricity, sharp knives) but they have different meanings and generate different levels of risk depending on their context and the individual who is confronted by the hazard. Risk is the likelihood that harm from the hazard will be realised (Wharton 1995). Risk is concerned therefore with probability or chance and the likelihood that danger or harm will result from exposure to a hazard" (p. 10).

There is evidence to suggest that there are differences in the nature and causes of both the crimes and accidents that visitors become victim to, or involved in. This view is supported by recent research for the Scottish Executive on tourist road accidents in rural areas, which concluded that:

> "Accident cause varies with the origin of the at-fault driver. Accidents caused by local drivers mostly involved loss of control, negotiating a bend, and going too fast. Accidents caused by foreign drivers involved driving on the wrong side of the road, turning and crossing the centre line. The principal causes of accidents due to U.K. visitor drivers were loss of control and over-taking. It seems clear from these data that foreign drivers are confused by having to drive on the left. The U.K. visitors perhaps lack skills in over-taking" (Sharples & Fletcher 2002: 39).

A feature which is certainly reiterating international research on tourist road accidents and the risks which visitors face in unfamiliar environments (Page *et al.* 2001).

In terms of crime there is also evidence to suggest that visitors are likely to be victimised (Wood Harper 2001) and subject to different types of crime than the local population (Barker *et al.* 2002); acts of terrorism have also become inextricably linked to tourism (Pizam & Smith 2000). Therefore, the visitor may actually face different types of hazards in addition to having different risk level from inherent hazards within the community environment.

Behaviour of the Visitor

There is a large body of literature on tourist behaviour almost as if the 'tourist' is something outside of normal society, a separate entity from the 'resident'. The 'tourist' is, in reality, one community member visiting another community. Yet in the transformation from 'resident' to 'tourist' there does seem to be evidence of behavioural changes in addition to the difficulties in adapting to an 'alien' environment that may have differing cultural values and expectations. Tarlow (2000) identifies six characteristics that make tourists more vulnerable to acts of crime:

- *Issues of trust* — the tourist may believe that his destination choice is safer than the normal place of residence resulting in behaviour that is more 'naïve' than the local residents;
- *Alienation from place* — the tourist will be unfamiliar with the area and may have few connections to it. They may also seek higher levels of adventure further separating them for the local populace;
- *Anomic behaviour* — the tourist may lack the usual social or ethical standards as they consume their 'holiday experience';
- *Cerebral hygiene* — as part of the holiday experience the tourist is often seeking to clear the mind and in doing so may relax normal defence mechanisms;
- *Lowering of inhibitions* — for many, being out with their normal environment releases them from their normal constraints and inhibitions, leaving them vulnerable to victimisation;

- *Stress* — in the desire to relax, there may be less care taken of valuable items with tourists often leaving articles such as bags or cameras in public places

These factors can equally be applied to accidents at destinations as they produce behaviours and situations leaving tourists vulnerable to accidents. These behavioural aspects, combined with the different risk levels identified earlier, can bring about undesirable consequences that the tourist cannot easily deal with due to lack of local knowledge.

Communication Channels

The idea of being safe and secure is fundamental to the idea of a modern society and there are high levels of expectations with regard to the facilities in place to ensure safety and security in our own areas. Not only do we look to organisations such as the police force to deliver a secure environment, we also use our own knowledge and experience of the area to take appropriate action to reduce risk. This may still be assessed badly due to incomplete information or due to a misunderstanding of the degree of risk attached to an activity. For example, the risk of road accidents is generally underestimated but the risk of being attacked by a complete stranger tends to be overestimated. However, as a general rule, as local residents in an area there is access to a wide range of formal sources (such as newspapers or local radio/TV alerts) and informal sources (such as neighbours' gossip or personal experience) that allow a broad assessment of potential risk to personal safety. Information on hazards in a local area, whether it is unsettled weather or criminal activity in a particular area, is often communicated through local media (newspapers, local radio and television) but also within informal communications such as discussion with friends or neighbours. These channels are rarely accessible to the visitor. The lack of information from these channels means that the visitor is less able to make informed choices. There is a need therefore to ensure that visitors receive relevant and up-to-date information on potential hazards through channels they can and will access despite the concerns of the destination and tourism industry to avoid negative publicity about the desirability of the area and locality. This point is argued very clearly in the chapter on tourism and midges, where there is a reluctance to reveal the scale and extent of the problem in the Highlands and Islands of Scotland (Blackwell & Page: ch. 12).

Although potential visitors may not have a detailed knowledge of an area, they will have a perception of the area possibly more based on image than reality. The image of the destination has a vital role to play both in the initial choice of destination made and in the subsequent attitudes, perception and behaviour of the visitor at the destination. The destination image can be described as:

> "an individual's overall perception or total set of impressions of a place,
> or as the mental portrayal of a destination" (Bigné, Sánchez & Sánchez
> 2001: 607).

As it based on individual perception, image is not objective but will vary depending on the messages received and how the potential visitor interprets these:

"destination image is formed by both stimulus factors and the tourists' characteristics" (Baloglu & McCleary 1999: 868).

In addition to influencing whether or not a visitor will choose to visit a particular destination, the image will also influence the visitor behaviour at destination (Bigné, Sánchez & Sánchez 2001: 608). As already argued, image will be a deciding factor in destination choice (Fry Bovet 1994; Baloglu & McCleary 1999); this is a relatively straightforward relationship in that a poor image, (i.e. one that implies danger to the tourist's well-being), is likely to result in a downturn in the number of visitors. However, once a destination is selected, then the image of the destination will still influence behaviour. This may be problematic on two counts, if the image is negative it may prevent them from fully participating in destination activities for example being afraid to venture out at night, which may reduce their spending, or they may have a very positive image that leads to behaviour that may put them at risk, such as not ensuring valuables are out of sight. This is certainly evident from the chapter on urban visitor safety in the book where Auckland was perceived as a safe locality to visit and yet tourist crime was a clear issue to address (see Barker *et al.* 2002: ch. 13). Visitors often behave on holiday in ways that they would not at home (for example, as the case of young people seeking 18–30 package holidays in Europe) and this can also apply to precautions and preventative measures that are neglected as a result of a 'safe' image combined with the holiday 'bubble' effect changing behaviour as discussed earlier in this chapter.

The image built up of a destination will come from many sources that are difficult to control but it is perhaps its most vital asset (Tilson & Stacks 1997). Studies have shown that news media is a major influence on the opinions of travellers (Fry Bovet 1994) and that news reports are much more likely to present a negatively biased report than positive or neutral (Wilks *et al.* 1996). Unfortunately there is often a 'ripple' effect when a crisis occurs that tarnishes the image of 'safe' destinations in other parts of a country or even neighbouring countries of an affected destination (Cavlek 2002) which has to be dealt with in a similar way to a crisis within the destination. A variety of complex factors are involved in the construction of a news media image (Avraham 2000) but its importance in influencing visitors to a destination suggests a need for careful consideration as to how any incidents are dealt with and presented to the news media.

As many of the chapters in this book emphasise, one of the challenges for the tourism is the number of disparate individuals and organisations that are involved in supplying and producing the finished product. Similarly the image is affected by many elements not all of which are directly controlled by the agencies involved in 'selling' the tourist product (e.g. September 11th terrorist attack and Foot and Mouth as two extreme examples). Often the image of a destination is affected not so much by actual events but how these events are reported in the media. The way that certain events are portrayed can influence the perceptions held of that destination. Without destination crisis management procedures and strategies in place to present a co-ordinated, coherent and unified response in the event of a major incident or series of incidents, then the destination image is likely to suffer.

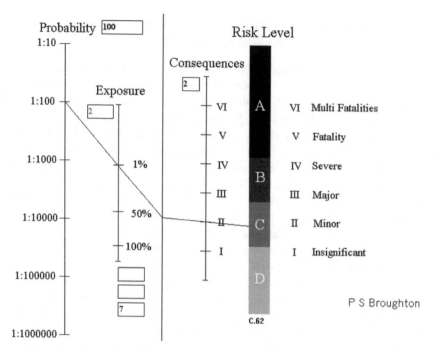

Figure 14.4: Risk assessment program.
(Copyright Paul Broughton, reproduced with permission)

Assessing Risk

The first stage in any attempt to manage risk must be to identify potential risk (Tchankova 2002) and then assess its likely impact in terms of probability of occurrence and consequences in the event of it actually occurring. Assessing risk level for a given incident can be achieved through risk assessment programmes such as the one illustrated in Figure 14.4. A calculation is made between the probabilities of an incident occurring, the exposure time of that incident occurring (in weeks, days or hours) together with the likely consequences of such an incident (e.g. minor consequences, fatalities or multi-fatalities). This produces a scale on which to measure incidents against each other and can allow an evaluation on how to allocate resources to optimum effect. This is a relatively easy exercise for an individual organisation but more problematic for a destination covering a variety of businesses and environments. However, as the destination is a contained area assessment should be entirely possible.

Once the extent of risk is ascertained, then the appropriate action must be decided upon. The Risk Evaluation Matrix discussed in Chapter 1 can be used to categorise risk by frequency and severity to allow decisions to be made on action to be taken. If there is both a high frequency of occurrence and a high severity of consequence then action may be taken to *avoid* the activity. In destination terms it may be to discourage visitors from going to certain areas of a town at night where there are regular problems of criminal activity. If on the other hand, frequency is high but the severity low, steps may be taken to *reduce* the risk. For example, after the appointment of new medical officers in Benidorm in the early 1990s better food hygiene measures were introduced that reduced incidents of diarrhoea from over 15% of British holidaymakers to less than 6% (Ryan 1996). If the severity of incidents is likely to be high but the frequency of occurrence is low then it may be appropriate to *transfer* the risk. In terms of an organisation this can be as simple as insuring against incidents causing injury or distress. However in destination terms, insurance could be having systems in place to deal with incidents that may occur such as the successful victim support services put in place for visitors to Ireland (Victim Support 2002). Risk is an inherent part of life and in some cases where frequency is low and the severity is low, risk may just be accepted as part of the characteristics of the destination for example if there are occasional 'breach of the peace' style problems where offence may be taken but no damage is done, then this may be accepted as reasonable to *retain*.

Acceptance: Should we 'Look After' the Visitor?

Ethics, Legalities and Economics

Health risks or fears such as the plague scare in India in 1994 (Grabowski & Chatterjee); or the terrorist campaigns aimed at tourists in Egypt in the early 1990s (Economist 1994); or fatal accidents involving tourists such as coach crashes in South Africa (Starrs 1999) or a series of incidents in the 'adventure' tourism business (Greenaway 1996) all have had significant impacts on the popularity of a destination. However, even less dramatic events or incidents can have an impact and concerns for safety, security or health and their subsequent portrayal in the media can dissuade tourists from visiting a particular locality for example a series of tourist mugging (African Eye News Services 2001). The safety of a destination is a major influence on holiday decisions. Brunt *et al.* (2000) in their study of victimisation and fear of crime stated that 53% of respondents in their study identified 'safe location' as an influence on holiday decisions. This bears out the findings presented in 1994 from the Burson-Marsteller survey which stated that concerns for safety in selecting a destination were rising and that for both U.K. and U.S. citizens naming "crime as a major concern followed by airline safety, terrorist attacks and access to medical facilities" (Fry Bovet 1994: 8). This concern is likely to have increased in the light of recent world events.

Having ascertained that visitors are subject to different degrees of risk, are likely to behave in a different manner to the local population, and that they do not have access to the same level of information as the local population, a decision has to be made on whether the destination should 'look after' visitors to their area. This may well depend on the degree to which the visitor is welcome in the area. Ethically, businesses within the tourism industry must have a duty to look after their customers similar to any business. However there can be an argument made that the destination has an even greater moral obligation to ensuring the safety and security of its visitors due to the image sold by the destination of an experience. Much of the literature on ethics in tourism focuses on service delivery and issues such as environmental impact, marketing, sustainability of development and education (Hultsman 1995) with little, if any, focus on the visitor; this despite the visitor being the purchaser of such services. Although it is recognised that tourism may have particular issues in terms of business ethics (Walle 1995), it seems strange that there is little apparent emphasis on the consumer of the product. If the visitor is of significant economic value there are hardnosed financial based reasons, in addition to any moral obligations, to reduce the potential risk to the tourist.

Tourism is now considered a vital aspect of the global economy and is favoured as an economic development option due to its relatively low initial investment requirements and the benefits it can bring to peripheral areas where other investment options may be difficult or not be viable, such as the Highlands of Scotland. It is clear that any loss to income from this source will damage local economies. Tourism can be affected by many different variables, some of which are outside of the control of the destination wishing to attract this income, such as the recent foot and mouth crisis or the terrorist attacks on America and the resulting hostilities. However, there are variables that can be controlled and systems put in place to encourage a sustainable tourism sector. Health, security and safety have been identified as particular areas of concerns that may affect the travel decisions and subsequent behaviour of visitors to a destination. Understanding the basic concerns and minimising risk is therefore in the interest of the destination community who may be dependent on tourism/visitor income or be concerned with an increase, or perceived increase, in crime in the local area and its cost to the local community both economically and socially.

Legal considerations may also affect the degree to which a destination may feel obligated to 'look after' visitor well-being. There has been a general increase in successful litigation recently that is likely to continue. Cases such as that of a swimmer in Bondi Beach in 1997 who successfully sued the local council after he had injured his head diving into shallow water in a flagged and patrolled area of water (Chipperfield 2002) have highlighted the tendency of courts to allocate responsibility and award substantial damages.

In part due to this increase in litigation but also due to world events, there has been a move by insurance companies to increase premiums and decrease coverage for certain activities affecting the tourism product. Since September 11th, there has been much news coverage on the issue of insuring against further terrorist attacks particularly for airlines and the combination of virtual 'uninsurability' and a decrease in numbers flying proved a fatal blow for many already struggling airlines. In some cases government subsidies were introduced to counteract this difficult situation (Economist 2001) and prevent further economic downturn from airline closures. However, at the destination there have also been changes and increases in insurance that predates September 11th for operators in the 'adventure' tourism market and event organisers (Chipperfield 2002). Given the level of increase of premiums and the increase in visitors wishing to participate in activities, this presents serious problems for the destination as a whole if operators are forced out of business or risk either underinsuring or not insuring. While the operator of the 'adventure' facility or the event organisers can quite clearly be seen as responsible for the visitor while they undertake activities, the implication to the destination if these facilities are not provided is likely to be a reduction in overall numbers. There may be a case for intervention on the part of the destination to counteract this problem in order to ensure continued provision of activities within the destination.

Having established the basic premise that the tourist requires special treatment at the destination in terms of information and/or precautionary measures on their behalf, who will provide this service? The responsibility on accommodation providers or attraction operators for example to provide a safe environment for their visitors is a 'duty of care' widely accepted throughout the world, however the wider responsibilities for warning visitors to a destination of potential hazards, the 'duty to warn' are less well established. The failure of those at destination level to develop systems to protect the tourist/visitor can have serious implications in terms of bad publicity and potential litigation. Traditionally the tourism industry has avoided dealing with the issue of warning visitors of potential dangers in order to avoid drawing attention to negative aspects of the destination; pointing out hazards and potential dangers can be seen as detrimental to the destination image and may be off putting to the visitor and either prevent their visit or reduce their spending as they avoid certain activities.

An incident occurring is likely to affect the visitor's perception of a destination but the way it is dealt with may have a stronger influence on the overall impression left in the visitor's mind. As this quote from Dame Helen Reeves, Chief Executive of Victim Support, expresses:

> "Becoming a victim of crime is a bad experience at any time. But to have
> it happen miles from home in a country where you do not know the law
> or the language can be devastating" (Victim Support 2002).

This is just as true for accidents as it is for crime. To be ill, injured, or victimised away from your normal environment and distanced from the familiarity of your usual support mechanisms could make even a relatively minor incident difficult to deal with, particularly if it is difficult to locate the necessary assistance due to lack of knowledge of the local systems. This would attest to a need to develop systems to not only protect visitors but also to provide necessary support should an incident occur. It would be naïve to assume that the existing safety and security measures that are in place for the local residents would be sufficient in terms of resources or strategies to deal with an influx of visitors that have different needs from the resident population.

Allocation: Who Should 'Look After' the Visitor?

Roles and Responsibilities

If the decision is reached by a destination that they have an obligation to the health and safety of visitors, they must decide on what is required to ensure visitor well-being. This can be achieved in two ways either by warning tourists of the potential hazards to equip them with the ability to make informed choices and/or to remove or reduce potential hazards. Given the earlier discussion on hazards and risks, it would be naïve to assume that the services and facilities provided for the safety and security for local residents would be sufficient to effectively deal with the safety and security issues brought about by an influx of visitors. The tourism facility and service providers could be seen as the obvious organisations to take responsibility for their clients safety and security, however the fragmented nature of the tourism product would be unlikely to allow a comprehensive approach to visitor safety in any but the 'custom' or 'enclave' destinations (see Figure 14.3) where one organisation is responsible for providing all facilities and services for the visitor. Destinations are not homogeneous and for each destination there will not only be differing health and safety requirements, there will also be differing resources available to meet these requirements.

Development of partnerships to ensure visitor well-being should be modelled on existing structures and mechanisms where these have been proven as effective in other areas of community well-being. Partnerships, community projects, consultations, co-operation and a whole host of similar buzz words have been the 'ideals' of the late 1990s, there has been an attempt in all areas of society to avoid the top down approach to all aspects of social control including policing. As ideals, the multi-agency, community-involved partnerships to crime prevention are laudable but their reality is based, not on empirical evidence of their usefulness in crime prevention in the community but on the political ideologies currently popular (Goris & Walters 1999). However examples from health promotion literature would suggest that such approaches can and do work but are most effective where there is local community involvement in setting agendas for action (Gillies 1998). Therefore care must be taken to develop the most effective collaborations. There may be a role for other, perhaps less obvious, organisations, for example Ryan (1996) identified possibilities of accessing information held by insurance companies, resulting in a mutually beneficial partnership, allowing patterns of incidents to be more easily identified and mechanisms to reduce such incidents to be put in place thus reducing claims and improving visitor well-being simultaneously.

There are other organisations involved in bringing visitors to any area that experiences significant visitor numbers and these organisations may be expected to share at least a degree of responsibility. According to Cavlek (2002) the tour operator has two key roles in terms of responsibility for the visitor:

• The operation or service procurement of various holiday components at the destination and the duty of care inherent in the provision of those components;
• Creating the image of the destination (Cavlek 2002).

However, although tour operators have an obligation to follow government advisory information with regard to crises, where events are at a less critical stage the stance taken by tour operators may be influenced by the level of investment they have in a particular destination (Cavlek 2002). There is an argument that the tour operator 'merely packages the tour' and therefore should not be held liable (Abbott & Abbott 1997). However, the tour operator often has influence and power in a destination, especially where it controls a large percentage of the bed-space. Travel agents can also influence the destination choice of travellers (Fry Bovet 1994) and may play a vital role in dissemination of information to potential visitors of a destination.

Often destination initiatives on safety issues are reactive. The case of Miami in 1993, after a spate of violent attacks against tourists, is typical of safety issues being addressed to counteract bad publicity. In an effort to reduce opportunistic attacks on tourists a partnership was formed between the community, the local convention and visitor bureau, policing services and civic leaders. Various steps were taken including increased signage, more police, better lighting and safety leaflets for distribution to tourists together with increased legislation and publicity for the safety measures undertaken (Eisman 1993). While this is a good example of groups working together for the benefit of the destination image and the safety of the visitor, it is unfortunate that these measures were not in place prior to the attacks as preventative measures.

The responsibility for aspects of visitor/tourist health and safety are likely to change depending on the perspective taken. People tend to judge behaviour and responsibility differently depending on their role within the equation. Early indications from current research being undertaken by one of the authors would suggest that there are distinct differences between visitor perception of who is responsible for them during their stay in a destination and the perceptions of those in industry (accommodation providers in this case). The visitors interviewed within a destination were more inclined to hold themselves responsible for their safety and security at destination than the accommodation providers did. This may have significance for strategies to improve visitor safety and security which may require focusing more on the empowerment of visitors than developing other 'protective' measures.

Conclusion

The safety of a destination has become an ever-increasing factor in the tourist decision-making model. However, there is little that the tourism industry on its own can do to ensure the safety of its client base at destination. This will not prevent operators from being held liable in the case of incidents and there has been increasing onus placed on individual component operators to not only fulfil their 'duty of care' but also to warn their clients of dangers in the destination as a whole. This 'duty to warn' does not fit easily with the carefree holiday market and does little for the overall image of the resort. Traditionally, responsibility would be abdicated to the police for the security of visitors to a destination, however, in isolation the police cannot effectively reduce incidents within existing resourcing. There has been recognition at all levels that individual agencies working on their own are not an effective use of resources. The development of partnerships can enhance service by developing an appreciation of the issues faced by each party involved and an understanding of their mode of operation. An effective partnership can also avoid duplication of effort and enjoy improved standing in the community; it can allow a proactive approach to ensuring visitor well-being and presenting a coherent, co-ordinated message to the media, particularly in times of crisis. Despite the obvious benefits of developing partnerships, there are inherent difficulties in bringing together diverse organisations with varying agendas. A multi-agency approach can flounder without a clear, practical focus. However, there are examples of 'best practice' that demonstrate the rewards to the destination of an effective incident reduction partnership. The experience of the tourist is clearly affected by events that affect their well being such as crime or ill health and also the way they are treated in the event of such incidents. This is an issue in terms of return visits and image of a destination but it also has implications for the providers of services designed to cope with health and safety provision as well as providers of tourism facilities. In these days of increased accountability and litigation, there is a need to appreciate and understand the issues affecting the visitor at destination level and to establish criteria for measuring visitor well-being and assessing responsibility for the overall well-being of the individual at a destination.

References

Abbott, J., & Abbott, S. M. (1997). Minimizing tour operators' exposure to lawsuits, *Cornell Hotel and Restaurant Quarterly*, (April), 20–24.

African Eye News Services (2001). South Africa: Two more tourists mugged at crime hotspot. *African Eye News Services*, (22/11/01).

Avraham, E. (2000). Cities and their news media images. *Cities, 17* (5), 363–370.

Baloglu, S., & McCleary, K. W. (1999). A model of destination image formation. *Annals of Tourism Research, 26* (4), 868–897.

Barker, M., & Page, S. J. (2002). Visitor safety in urban environments: The case of Auckland, New Zealand. *Cities: The International Journal of Urban Policy and Planning, 19* (4), 273–282.

Barker, M., Page, S. J., & Meyer, D. (2002). Modelling tourism crime. *Annals of Tourism Research, 29* (3), 762–782.

Bentley, T., & Page, S. J. (2001). Scoping the extent of tourist accidents in New Zealand. *Annals of Tourism Research, 28* (3), 705–726.

Bigné, J. E., Sănchez, M. I., & Sănchez, J. (2001). Tourism image, evaluation variables; and after purchase behaviour: inter-relationship. *Tourism Management, 22*, 607–616.

Brunt, P., Mawby, R., & Hambly, Z. (2000). Tourist victimisation and the fear of crime on holiday. *Tourism Management, 21*, 417–424.

Cavlek, N. (2002). Tour operators and destination safety. *Annals of Tourism Research, 29* (2), 478–496.

Chipperfield, M. (2002). Threat to sports holidays. *The Daily Telegraph*, (22/06/02).

Economist (1994). Staying away: Egypt. *Economist, 330, 7851*, 45–46.

Economist (2001). Is terrorism an uninsurable risk? *Economist*, (12/11/01), 21–24.

Eisman, R. (1993). Violence against tourists forces Miami to address safety issues. *Public Relations Journal, 49* (12), 10–12.

Fry Bovet, S. (1994). Safety concerns world travel market. *Public Relations Journal, 50* (3), 8.

Gillies, P. (1998). Effectiveness of alliances and partnerships for health promotion. *Health Promotion International, 13* (2), 99–120.

Goris, P., & Walters, R. (1999). Locally oriented crime prevention and the "partnership approach". *Policing: an International Journal of Police Strategies & Management, 22* (4), 633–645.

Grant, M. (2002). *Local well-being: Re-engineering tourism on the Isle of Wight.* Unpublished conference paper presented at the Tourism Research 2002 Conference in Cardiff.

Greenaway, R. (1996). Thrilling not killing — managing the risk tourism business. *Management*, (May), 46–49.

Hall, C. M., & Page, S. J. (2002). *The geography of tourism and recreation: Environment, place and space* (2nd ed.). London: Routledge.

Hultsman, J. (1995). Just tourism: An ethical framework. *Annals of Tourism Research, 22* (3), 553–567.

Massam, B. H. (2002). Quality of life: public planning and private living. *Progress in Planning, 58,* 141–227.

Pacione, M., & Gordon, G. (1984). *Quality of life and human welfare.* London: Royal Geographical Society.

Page, S. J. (1997). *The cost of accidents in the adventure tourism industry.* Consultants Report for the Tourism Policy Group. Wellington: Ministry of Commerce.

Page, S. J. (2002). Tourist health and safety. *Travel and Tourism Analyst,* in press.

Page, S. J., & Hall, C. M. (2002). *Managing urban tourism.* Harlow: Pearson Education.

Page, S. J., & Meyer, D. (1996). Tourist accidents: an exploratory analysis. *Annals of Tourism Research, 23* (3), 666–690.

Page, S. J., Bentley, T., Meyer, D., & Chalmers, D. (2001). Scoping the extent of road safety: Motor vehicle traffic accidents in New Zealand 1982–1996. *Current Issues in Tourism, 4* (6), 503–526.

Pizam, A., & Smith, G. (2000). Tourism and terrorism: A quantitative analysis of major terrorist acts and their impact on tourism destinations. *Tourism Economics, 6* (2), 123–138.

Ryan, C. (1996). Linkages between holiday travel risk and insurance claims: Evidence from New Zealand. *Tourism Management, 17* (8), 593–601.

Sharp, B. (2001). *Strategies for improving mountain safety: Report of a research study.* The Leverhulme Trust/University of Strathclyde.

Sharples, J. M., & Fletcher, J. P. (2001). *Tourist road accidents in rural Scotland.* Edinburgh: Scottish Executive Central Research Unit.

Smith, D. M. (1977). *Human geography: A welfare approach.* London: Edward Arnold.

Starrs, C. (1999). Untitled. *Herald,* (28/09/99), page 3.

Tarlow, P. E. (2000). Creating safe and secure communities in economically challenging times. *Tourism Economics, 6* (2), 139–149.

Tchankova, L. (2002). Risk identification — basic stage in risk management. *Environmental Management and Health, 13* (3), 290–297.

Tilson, D. J., & Stacks, D. W. (1997). To know us is to love us: The public relations campaign to sell a 'business-tourist-friendly' Miami. *Public Relations Review, 23* (2), 95–115.

Victim Support (2002). New rights for victims of crime in Europe: Council Framework Decision on the standing of victims in criminal proceedings. *Victim Support,* (June), 15–17.

Walle, A. H. (1995). Business ethics and tourism: From micro to macro perspectives. *Tourism Management, 16* (4), 263–268.

Wilks, J., Pendergast, D., & Service, M. (1996). Newspaper reporting of tourist health and safety issues. *Australia Leisure,* (September), 45–48.

Wood Harper D. Jr. (2001). Comparing tourists crime victimization. *Annals of Tourism Research, 28* (4), 1053–1056.

Author Index

Subject Index